STRATOSPHERIC OZONE

United Kingdom
Stratospheric Ozone
Review Group

First Report

Prepared at the request of the
Department of the Environment and the Meteorological Office

LONDON: HER MAJESTY'S STATIONERY OFFICE

First published 1987
ISBN 0 11 752018 7

United Kingdom
Stratospheric Ozone Review Group

Dr A F Tuck* *(Chairman)*	Meteorological Office, Bracknell
Dr R A Cox	Harwell Laboratory
Dr J Farman	British Antarctic Survey, Cambridge
Dr L J Gray	Rutherford Appleton Laboratory, Chilton
Dr R L Jones	Meteorological Office, Bracknell
Dr A N O'Neill	Meteorological Office, Bracknell
Dr S A Penkett	University of East Anglia, Norwich
Dr J A Pyle	University of Cambridge
Dr H K Roscoe	University of Oxford
Dr J Hollies *(Observer)*	ICI plc, Runcorn
Dr G J Jenkins *(Executive Secretary)*	Department of the Environment, London

* Dr Tuck took up a post at NOAA, Boulder, in July 1986.

Contents

EXECUTIVE SUMMARY

Ozone in the atmosphere is important to man because it reduces the amount of solar ultra-violet radiation reaching the ground; this radiation has the potential to affect adversely the environment and human health. The amount of UV radiation absorbed depends on the amount of ozone throughout the depth of the atmosphere ("column ozone"). Concentrations of ozone in the atmosphere are determined by a complex set of processes: the circulation of the atmosphere, the chemical reactions that take place in it, and the effects of radiation.

Statistical techniques show that there is no significant trend detectable in measurements of the global average of column ozone. However, there are large variations on many time-scales which would make a small trend (say, less than 1% per decade) difficult to identify. Ozone in the upper stratosphere appears to have decreased by about 2-3% since 1978, although measurement of this quantity is particularly error-prone.

Recently a large depletion, about 40%, has been seen in stratospheric ozone over Antarctica in the spring — the so-called "ozone hole". The cause is not yet known. It may be a natural change in atmospheric circulation; it may be chemical, due to increasing levels of chlorine and/or bromine from man's activities or it may involve both of these. We do not yet know if there are implications for ozone on a global scale. Large experimental campaigns mounted by the US in 1986 and 1987 may resolve these uncertainties.

The concentrations of several of the source gases which influence the amount of ozone in the atmosphere have been increasing for some years. Annual increases in methane (CH_4), nitrous oxide (N_2O) and carbon dioxide (CO_2) are about 1%, 0.25% and 0.4% of their present concentrations respectively. Emissions of many chlorofluorocarbons (CFCs) are now increasing again, following a period of decline from 1974 to 1982.

Complex mathematical models of the physical and chemical processes in the atmosphere are used in attempts to assess the effects of these changes in source gases on atmospheric ozone. They predict that increasing levels of CFCs and nitrous oxide will each give rise to a depletion of ozone in the stratosphere. Carbon dioxide and methane, on the other hand, will act to increase ozone; the increase due to methane will occur largely in the troposphere.

Models can reproduce many observed features of the distribution of trace gases in the stratosphere. On the other hand, comparisons of some specific features reveal inadequacies in our present understanding; this limits the confidence with which models can be used for predictive purposes. We have no firm basis for quantifying the uncertainty in the predictions.

To try to assess how ozone will change over the next 50 years or more, estimates of future source gas concentrations are required for numerical model calculations. Large uncertainties attach to these estimates, and to the model calculations. In the case of a continuing annual rise in concentrations of methane by 1% and N_2O by 0.25%, carbon dioxide growing by 0.5% per year and with CFC emissions assumed to grow at a rate of 3% per year, one dimensional (altitude only) models predict that globally-averaged column ozone will have decreased by about 1% by the year 2000, and by about 4% after 50 years. If CFC emissions continue at about present levels, with other source gas changes as above, then little ozone depletion will occur.

In contrast, recent two-dimensional (altitude, latitude) models predict overall larger decreases in globally-averaged column ozone than those predicted by the I-D models. Using the same scenario as above with CFCs assumed to grow at 3% per year, they predict a global depletion of 3% by the year 2000. Furthermore the 2-D models indicate that depletion is likely to be greater at high latitudes; at 60°N during spring, for example, it will be about twice as large as the global average. It should be noted that these models, incorporating only "conventional" chemical reactions, are unable to explain the very large, rapid, depletion of ozone occurring over Antarctica.

Even source gas scenarios for which models predict little or no change in column ozone may give rise to significant changes in the vertical distribution of ozone. These changes may affect the circulation of the stratosphere.

The lifetime of CFCs in the atmosphere is about 100 years. Any changes in stratospheric ozone brought about by CFCs will be long-lived. Model predictions show that in the hypothetical case of a complete cessation of CFC emissions ozone will continue to decline slightly for a further 5-10 years before recovery begins, but the additional depletion will be less than one percent. The overshoot will be greater, and the recovery time longer, if CFC emissions are reduced rather than terminated completely. These predictions depend critically on assumptions of changes in other source gases, particularly methane.

All the source gases which influence ozone concentrations are also greenhouse gases, as is ozone itself in the troposphere. Rising concentrations of these gases are expected to lead to an increase in the surface air temperature — the "global warming". Current models, assuming continued growth in concentrations of greenhouse gases for the next 50 years, estimate a global temperature rise of between 1.5 and 4.5°C. There are large uncertainties in this estimate. Moreover the realisation of this warming may be delayed by decades due to the thermal inertia of the oceans.

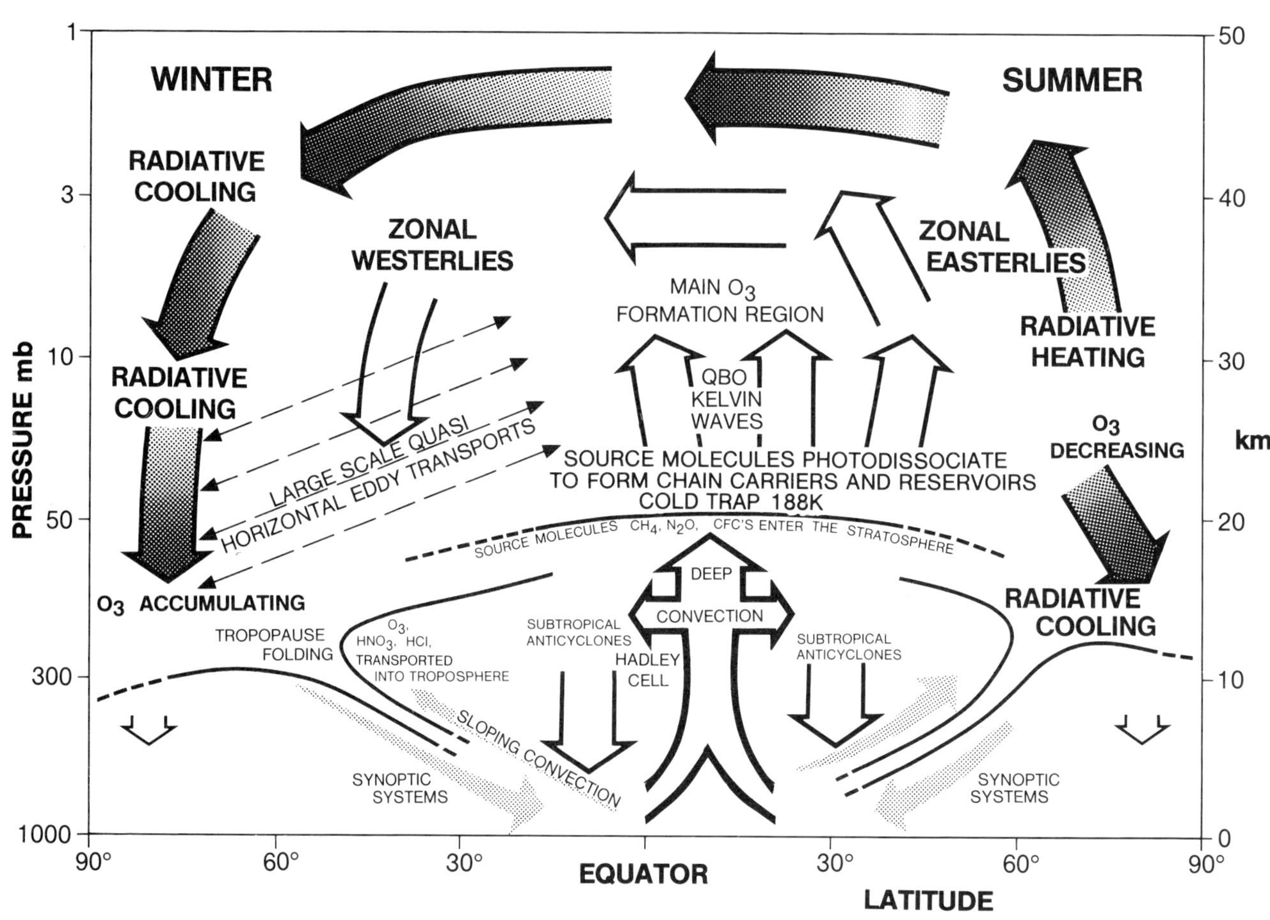

FRONTISPIECE

General arrangement of dynamical, radiative and chemical processes in the earth's atmosphere which determine the distribution of stratospheric ozone.

INTRODUCTION 1

BACKGROUND

The stratosphere is that region of the earth's atmosphere between about 10 km and 50 km above the ground. What is commonly called the ozone layer is the region in the stratosphere where ozone makes up a greater proportion of the air than at any other height in the earth's atmosphere. Although this proportion is only a few parts per million, the presence of the ozone is vital because it absorbs ultraviolet light and prevents too much of it from reaching the earth's surface where it has the potential to affect living things. The amount of ozone in the stratosphere is in dynamic balance between production and destruction, determined by a complex set of physical and chemical processes; the dynamics of the atmosphere (winds and turbulence), chemical reactions taking place within it, and the effects of short wave (solar) and long wave (terrestrial) radiation on the constituents. The gases which take part in these processes, and from which other important constituents are derived, are generated near the earth's surface and are referred to as source gases. Changes, either natural or man-made, in the production or loss of these source gases could perturb this balance, and lead to changes in the amount of ozone in the stratosphere.

Because nitrogen compounds play a major role in the chemistry of the stratosphere, it was the potential effect on ozone of fleets of high altitude supersonic aircraft (SSTs) which first gave rise to concern in 1971. In the UK, the Department of Trade and Industry and the Meteorological Office set up the Committee on the Meteorological Effects on Stratospheric Aviation (COMESA) in 1972 and, following extensive research at the Meteorological Office, the National Physical Laboratory and in universities, an assessment of the effects of SSTs was published in 1975 (COMESA 1975) which concluded that, using fleet sizes projected at that time, there was no real threat to stratospheric ozone. Other studies in the US and France came to similar conclusions.

By this time, the classic paper by Rowland and Molina had appeared in the literature. This postulated that long-lived man-made compounds containing chlorine (chlorofluorocarbons – CFCs) released from the earth's surface would, after transport to the stratosphere, be decomposed by sunlight into free chlorine which would increase the rate of destruction of ozone. This new theory led to research being pursued with renewed vigour on both sides of the Atlantic. In 1976, the Meteorological Office and the Department of the Environment formed the Stratospheric Research Advisory Committee (STRAC) to provide a comprehensive assessment, following an extensive co-ordinated research programme. This they did in Pollution Paper 15 (DOE, 1979) which concluded that continued CFC emissions at the 1975 rate would ultimately result in depletion of the ozone column of between 11% and 16%. It also concluded that these figures were very uncertain, and that some of the errors associated with the highly averaged one-dimensional and two-dimensional model calculations were unquantifiable.

During the early 1980s continuing research led to a steady increase in knowledge, and this allowed calculations of ozone layer depletion to be updated. The Coordinating Committee on the Ozone Layer, convened in 1977 by the UN Environment Programme (UNEP), continued to issue revisions of the Ozone Layer Bulletin (UNEP 1983, for example), in which research was reviewed and a "best guess" figure for forecasts of ozone layer depletion was given, all these estimates tended to be lower than the 1979 figures quoted above. During this period it was realised that any assessment of change in the total amount of ozone in a column above the earth's surface must include the effect, not only of changes in stratospheric ozone but also in ozone in the troposphere (ie below about 10 km), and that both of these, although not determined by entirely the same processes, depend critically on the future course of not only CFC emissions but of concentrations of other trace gases such as methane, carbon dioxide and nitrous oxide. Furthermore, changes in levels of each of these trace gases are expected to give rise to a global climate change, and this in turn could affect levels of ozone in the atmosphere.

In 1981 UNEP decided that the time had come to do more than periodically review the scientific situation, and set about organising a Convention for the Protection of the Ozone Layer which, following negotiations, was agreed at Vienna in March 1985. The Convention contains articles on the pursuit of research and on the exchange of information etc, and the UK ratified it on 15 May 1987. Further discussions are now taking place under the Convention to consider the need for, and the form of, any proposals to control the production or use of CFCs.

General interest in the ozone layer and the possible threat to it from man's activities has grown rapidly again since 1984 for two main reasons. Firstly, the negotiations towards a Convention and possible regulatory protocols have demanded the production of an authoritative statement of current scientific understanding, so that decisions may be based on sound science. Secondly, findings published in recent scientific papers have, rightly or wrongly, given cause for concern amongst the general public and policy-makers. In one such paper, the theory was put forward that ozone depletion may not always be proportional to chlorine injections into the atmosphere, but that a situation could be reached where ozone depletion would accelerate rapidly. Another recent paper showed clearly that amounts of ozone over Antarctica in September and October have decreased by up to 40% since the late 1970s.

In 1984 NASA (supported by UNEP, WMO, NOAA, FAA, BMFT and CEC) brought together leading scientists and asked them to prepare an authoritative assessment of our current understanding of the ozone layer and man's threat to it. This large task, involving about 175 scientists from 10 countries, was completed in early 1986 with the publication of three 500 page volumes which will remain the reference document for some years to come. About a dozen scientists from the UK contributed to this Assessment, and four of them each chaired one of the twelve chapters.

The NASA-WMO-UNEP Assessment (referred to in this report as NASA/WMO, 1986) is a large, reference document, but not one which will be readily understood by policymakers. In 1985 the Department of the Environment and the Meteorological Office asked a small number of UK scientists, drawn mostly from contributors to the large Assessment, to make up the Stratospheric Ozone Review Group which would prepare a short review to address the main scientific issues in a form more easily understood by generalists. The membership of the Group, and its Terms of Reference, are given in Annex A. This report is the result of their labours. Its scope is limited to consideration of the physical and chemical aspects of ozone depletion and the related subject of climate change. Possible effects on human health, on aquatic or terrestrial life, or on materials, due to increases in UV flux caused by ozone depletion, have been excluded as have potential effects of climate change. Furthermore, it does not include any discussion on economic factors, such as the estimation of future growth in CFC emissions or the ease with which they could be replaced by other chemicals. The opportunity has been taken to include new developments in our understanding of, or observation of, the ozone layer since the publication of the NASA-WMO-UNEP Assessment, especially with regard to the phenomenon of Antarctic ozone depletion.

THE STRUCTURE OF THE REPORT

This report seeks to summarise the complex physics and chemistry lying behind the cyclical process of emission of source gases; their dispersion and transport into the stratosphere, followed by their decomposition and the interaction of the reactive products with ozone.

Chapter 2 introduces the topic of ozone in the earth's atmosphere. Chapter 3 outlines the way in which the source gases are changing in concentration, and the relevance of these changes. Chapter 4 describes the detailed chemistry of the stratosphere, and Chapter 5 gives a summary of relevant atmospheric physics; radiation and dynamics. The morphology of ozone, spatial and temporal patterns and trends, are dealt with in Chapter 6. In Chapter 7 measurements of other gases in the stratosphere are considered; not in order to give a complete description but to point out the major areas of ignorance.

A brief description is given, in Chapter 8, of the way in which assessment models, which seek to predict changes in ozone, are formulated; this is followed in Chapter 9 by a resume of the results from these models. The large depletion in stratospheric ozone in springtime over Antarctica has been the most significant discovery of recent years, and understanding it is vital to the credibility of models used for the assessment of ozone depletion. For this reason a separate chapter, Chapter 10, has been devoted to the description of this phenomenon and to the hypotheses which have been put forward to explain it.

Chapter 11 deals with the whole topic of the greenhouse effect and climate change albeit in a very condensed form. In Chapter 12 a review is given of current activities in the United Kingdom related to stratospheric ozone, and finally in Chapter 13 there is a list of recommendations for further work. An Executive Summary is provided at the head of the report.

In the interests of brevity and clarity, and in consideration of the likely readership, the normal procedure of giving scientific credit by use of references has been suspended for the bulk of this report. The NASA-WMO-UNEP Assessment can be consulted for more detail in specific areas, and this has an extensive bibliography and reference list. Scientific references are given in this present report where they have been published since the Assessment.

OZONE IN THE EARTH'S ATMOSPHERE 2

THE EARTH'S ATMOSPHERE

The atmosphere of the earth is a thin layer of gases covering its surface; its density decreases as we ascend, so that about half of its mass is below 5 km. The temperature of the atmosphere behaves differently, as shown in Figure 2.1a. It usually decreases with height above the ground through a region known as the troposphere, until a height of about 10 km (the tropopause) where it starts to increase again.

The region above the troposphere where the temperature increases with altitude is known as the stratosphere. This increase of temperature with height means that vertical mixing (convection) is supressed; this distinguishes it from the troposphere where vertical mixing is rapid. The stratosphere extends to a height of about 50 km (the stratopause); only about 0.1% of the mass of the atmosphere is above the stratopause.

The bottom of the atmosphere is warm because it is close to the earth's surface, itself warmed by solar radiation. The source of heat in the stratosphere is the absorption of solar ultra-violet light by the ozone molecules present there — the ozone which is the subject of this report.

CONSTITUENTS OF THE ATMOSPHERE

The atmosphere has evolved over geological time into a mixture of approximately 80% nitrogen and 20% oxygen. It also contains many other gases at concentrations ranging from a few percent down to a few parts per million million (10^{12}). The biochemical, geochemical and geophysical importance of these "trace" gases is much greater than their relatively low concentrations would suggest. This is particularly true of ozone and a number of other trace gases ("source gases") which control the abundance of ozone. In addition these gases, together with water and carbon dioxide, exert a profound influence on the earth's temperature through their effect on the transfer of radiation through the atmosphere.

Ozone is produced and destroyed at a wide range of altitudes in the atmosphere; in the polluted boundary layer (up to a few hundred metres) and in the free troposphere as well as in the stratosphere. The production and loss mechanisms are different in the different regions. This report will deal mainly with the stratosphere, a parallel report from the Photochemical Oxidant Review Group (PORG, 1987) deals with chemistry of the boundary layer and troposphere. It must be remembered, however, that the amount of ozone present through the whole depth of the atmosphere (referred to as "column ozone"), which controls the amount of ultra-violet radiation reaching the ground, is determined by production and loss at all levels, and when calculations of column ozone change are carried out, they include consideration of processes in the troposphere as well as the stratosphere.

As a result of chemical, radiative and dynamical processes, the ozone concentration varies throughout the atmosphere as shown in Figures 2.1b and 2.1c. In the latter, ozone is given as a mixing ratio, ie the proportion by volume in which it is present. In the former diagram it is shown in terms of its absolute density, the number of molecules per cubic centimetre. The shape of the ozone density profile, below about 25 km, is different at different latitudes and in the different seasons. The profile illustrated is a long period average at mid-latitudes. About 90% of all the ozone in the atmosphere resides in the stratosphere.

ATMOSPHERIC DYNAMICS AND TRANSPORT

Ozone, and other gases relevant to its photochemistry, are transported by circulations on a wide range of spatial and temporal scales. Ozone absorbs solar ultra-violet radiation, so the changing distribution of ozone alters the pattern of

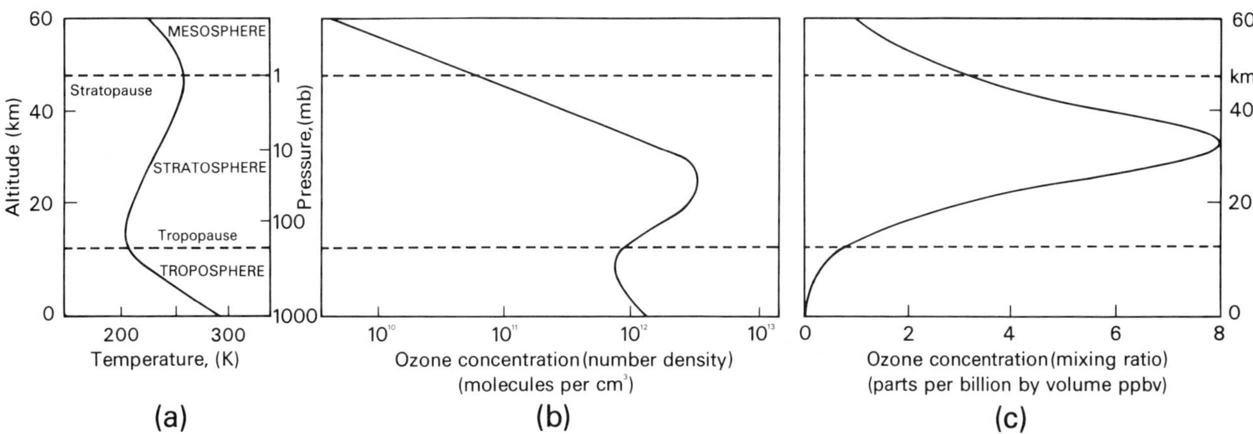

Figure 2.1 (a) The variation of temperature with altitude through the atmosphere
(b) The altitude profile of ozone number density (molecules per cubic centimetre)
(c) The altitude profile of ozone mixing ratio (parts per ~~billion by~~ volume — ~~ppbv~~)
million ppmv

radiative heating. Radiation affects transport because air parcels move systematically upwards and away from areas being heated into areas that are being cooled where they descend. Air is also made to move by perturbations that develop in the circulation. For example, in the winter stratosphere there are planetary-scale disturbances entering the stratosphere from the troposphere below. These also serve to drive the atmosphere away from radiative equilibrium. Hence, there is a strong correlation between radiation, dynamics and transport.

Small scales of motion are also essential in governing both the atmospheric circulation and the distribution of chemicals within it. There are turbulent fluxes of momentum, heat, moisture and trace gases at land and sea surfaces. Latent heat is released when clouds form, and the clouds affect the transfer of radiation. Deep convective cells at equatorial latitudes deposit water vapour in the lower stratosphere, and tongues of stratospheric air descend along fronts in the upper troposphere at middle latitudes. Gravity waves, generated by mountains for instance, propagate upwards and eventually degenerate into turbulence which alters the large-scale flow and associated transport.

ATMOSPHERIC CHEMISTRY

Ozone (O_3) is produced directly in the stratosphere by the action of ultra-violet light ($h\nu$) with a wavelength of less than 243 nm on oxygen molecules (O_2), which then photodissociate to form free oxygen atoms (O). These react with further oxygen molecules to form ozone:

$$O_2 + h\nu \rightarrow O + O \qquad \lambda < 243 \text{ nm} \qquad (1)$$

$$O + O_2 + M \rightarrow O_3 + M \qquad (2)$$

(M is any other molecule)

Ozone itself is photodissociated by both UV and visible light:

$$O_3 + h\nu \rightarrow O_2 + O \qquad (3)$$

The production processes (1) and (2) are balanced by loss due to chemical reactions. Until the 1950s, chemical loss of odd oxygen (ie O and O_3) was attributed only to the reaction:

$$O + O_3 \rightarrow O_2 + O_2 \qquad (4)$$

originally proposed by Chapman in 1930. It is now realised that ozone in the stratosphere is removed predominantly by catalytic chain reactions (ie reactions in which the molecule initiating the reaction sequence is regenerated and is therefore able to initiate further reactions) involving reactive molecules,

$$X + O_3 \rightarrow XO + O_2 \qquad (5)$$

$$XO + O \rightarrow X + O_2 \qquad (6)$$

$$\text{net: } O + O_3 \rightarrow 2O_2$$

where the catalyst X = H, OH, NO, Cl and Br. Thus these species can, with varying degrees of efficiency, control the abundance and distribution of ozone in the stratosphere. Assignment of the relative importance, and the prediction of the future impact, of these catalytic species depends on a detailed understanding of the chemical reactions which form, remove and interconvert the active components of each family. This in turn requires knowledge of the atmospheric life cycles of the hydrogen, nitrogen and halogen-containing source and sink molecules, which control the overall abundance of these reactive molecules.

The presence in the atmosphere of the trace gaseous constituents having an influence on the chemistry of stratospheric ozone, results from natural or man-made emissions, either in the biosphere, lithosphere or hydrosphere (oceans). Compounds produced by life forms (including man) in the biosphere have a dominant role.

A common feature of these so-called "source gases" is that they are relatively (sometimes completely) inert in the troposphere, but are transported up to the high stratosphere where they are broken down to produce the reactive chemicals which act as the catalysts for ozone destruction. Ozone, which is present in the atmosphere at concentrations of a few parts per million, is controlled by other gases whose concentrations rarely exceed 10 parts per billion (1 part in 10^8) by volume (ppbv). This control is exerted by virtue of the fact that the catalyst species can initiate up to several thousand ozone-destroying cycles (reactions 5 and 6) before they are themselves removed.

The source of NO_x (NO and NO_2) in the stratosphere, for example, is mainly N_2O. The source of chlorine species (ClO_x) in the stratosphere are methyl chloride (CH_3Cl), carbon tetrachloride (CCl_4) and the chlorofluorocarbons (CFCs). The source of the hydrogen (HO_x) species is methane and water vapour. These source gases are all emitted into the troposphere, and many of them (for example N_2O, CFCs) are relatively inert there. The removal of active chemicals from the stratosphere occurs by chemical conversion to more stable "sink" molecules, eg nitric acid (HNO_3), water and hydrochloric acid (HCl), which are transported down into the troposphere and ultimately removed by rain.

The reactions described in this background section are only the very simplest ones; in computer models which attempt to simulate the behaviour of the real atmosphere, from 50 to 200 or more thermal and photochemical reactions are required in order to provide an adequate description of the chemistry governing production and loss of ozone and the relevant trace gas families.

MODELS OF THE ATMOSPHERE

Because of the complexity of physical and chemical processes that determine the distribution of ozone in the atmosphere, the only way to predict the effect on ozone of changes in the concentrations of the various source gases is to use computer based mathematical models, in which these processes are represented. These computer models may be one-dimensional, giving globally averaged ozone as a function of altitude only, or two-dimensional in which the latitude dependence is described. So far no three-dimensional (altitude, latitude, longitude) General Circulation Models (GCMs) have incorporated chemical reactions for use in perturbation studies. Computer models can only be used for prediction once confidence in them has been established. Comparing the model calculations of present day concentrations, (not only of ozone, but of other chemical species) with measurements, will allow us to make some tests of the model, but so far these tests have not been comprehensive, largely because of a shortage of adequate observations.

Uncertainty in model predictions of ozone changes in the future arise from several causes. The predictive models themselves may be deficient, because known physical or chemical processes are not adequately represented. Models have to be simplified to a certain extent so that they can be run on existing computers. To do this, parametrization is used, whereby quantities, physical or chemical, which are not explicitly represented are expressed in terms of those that are. The theoretical basis for this is often far from satisfactory.

Models may also be deficient because they miss out significant atmospheric processes, of which we are simply unaware. There may be chemical reactions, for instance, which have a profound effect on ozone concentrations, but which have not yet been incorporated into models of the atmosphere. Lastly, the model calculations of ozone changes in the future are driven by assumed future scenarios of source gas concentrations, and these may turn out to be wrong. In the case of methane and nitrous oxide, for instance, we assume a continuation of present trends, because we lack an adequate understanding of the reasons for these changes. In the case of CFCs, because the emissions are totally under man's control, a range of possible emission increases is assumed.

THE IMPORTANCE OF OZONE

Ozone in the atmosphere is important for a number of reasons. Firstly, ozone absorbs solar ultra-violet light, light which has the potential to adversely affect the environment (aqueous and terrestrial ecosystems, materials) and human health (the incidence of certain types of skin cancer), though there are large uncertainties with regard to dose-response relationships. The total amount of ozone in the whole depth of the atmosphere per unit area (column ozone) determines how much of this ultraviolet light reaches ground level, and any change in this ozone column could have significant consequences. Secondly, as we have already mentioned, when ozone absorbs ultra-violet light it heats the atmosphere in the vicinity; a change in the abundance of ozone may change the temperature structure of the atmosphere and thereby alter the atmospheric circulation.

Thirdly, ozone in the troposphere is a greenhouse gas. Like carbon dioxide and other source gases (methane, nitrous oxide, CFCs) it is relatively transparent to solar radiation (except in the ultra-violet, as discussed above) but absorbs strongly in the infra-red part of the spectrum. The surface temperature of the earth is the result of a balance between incoming solar radiation and outgoing infra-red radiation emitted by the earth. Hence any additional absorption of the latter will trap heat and casue the earth to become warmer by the greenhouse effect. Increased concentrations in any of these greenhouse gases have the potential to cause a change in the earth's climate. Lastly, ozone at ground level can itself have a direct effect on the environment and, in extreme cases, human health. It may be implicated in damage to trees which has recently been observed in Europe and North America.

SOURCE GASES 3

During the last decade and a half, measurements have shown that a large range of trace gases emitted at the earth's surface dissociate in the stratosphere to produce highly reactive chemical species which can have a profound influence on the earth's ozone layer. They are thus the source of the atoms and free radicals which react with ozone and are therefore termed source gases.

HALOGENS

Many individual source gases containing chlorine and/or fluorine are present in the atmosphere; at least twenty have been identified positively so far. Only three compounds, CH_3Cl (methyl chloride), CH_2Cl_2 and $CHCl_3$ are thought to have substantial natural sources and these represent the predominant forms of natural organic chlorine compounds in the atmosphere. It was thought for a while that carbon tetrachloride (CCl_4) also had a significant natural source, but this is probably not the case. The major anthropogenic source gases for stratospheric chlorine to date are in the form of $CFCl_3$ (CFC 11) and CF_2Cl_2 (CFC 12) and CCl_4 with increasing contributions from methylchloroform (CH_3CCl_3), CHF_2Cl (CFC 22) and $CF_2Cl.CFCl_2$ (CFC 113) (see Table 3.1).

Cape Grim,Tasmania

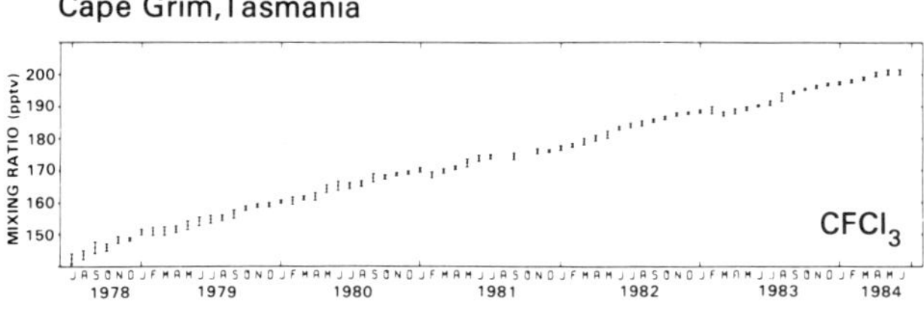

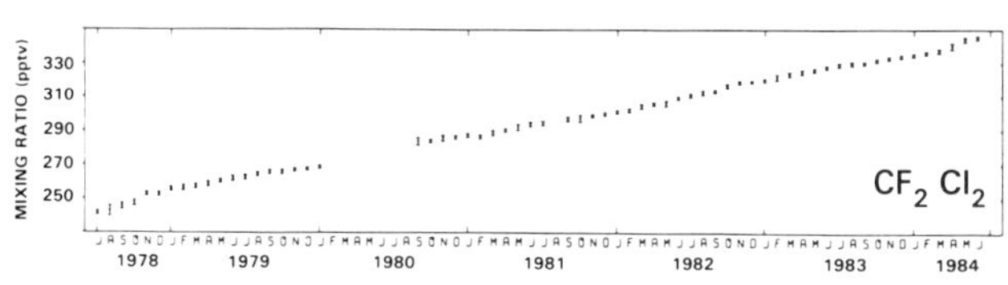

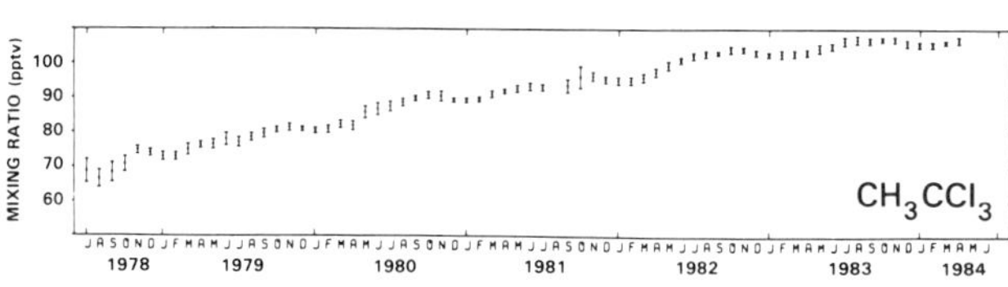

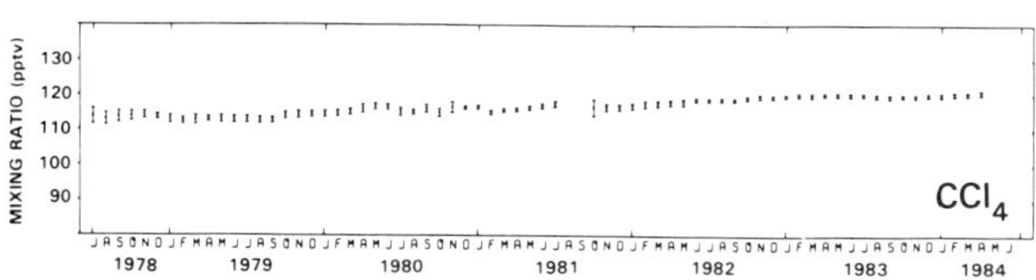

Figure 3.1 Trends in source gases at the ALE/GAGE station at Cape Grim, Tasmania. Values shown are monthly mean mixing ratios from observations made 4 times a day; the bars represent the monthly variance

The total organic chlorine content of the troposphere, which acts as a source for chlorine in the stratosphere, is in excess of 2.8 ppbv of which at least 2.2 ppbv results from man-made emissions. Consequently the majority of chlorine in the chlorine-ozone removal cycle shown below is man-made.

Chlorine atoms are released from their source gases either by photolysis as, for example, in the case of $CFCl_3$:

$$CFCl_3 + h\nu \rightarrow CFCl_2 + Cl \qquad (7)$$

or by reaction with free radicals, as, for example, in the case of CH_3Cl

$$CH_3Cl + OH \rightarrow Cl + \text{products} \quad \text{(multistep)} \qquad (8)$$

The chlorine atoms then react with ozone in a catalytic cycle which effectively removes two molecules of ozone without consuming chlorine:

$$Cl + O_3 \rightarrow ClO + O_2 \qquad (9)$$
$$ClO + O \rightarrow Cl + O_2 \qquad (10)$$

Many bromine and iodine compounds have predominantly natural sources. This is probably the case for $CHBr_3$, CH_2Br_2, CH_2BrCl, $CHBr_2Cl$ and CH_3I. However, CF_3Br (Halon 1301), CF_2BrCl (Halon 1211), $C_2H_4Br_2$, are entirely man-made, and it is likely that the man-made emissions of CH_3Br (methyl bromide) are predominant. Thus the importance of the bromine-ozone removal cycle is strongly influenced by human activities.

$$Br + O_3 \rightarrow BrO + O_2 \qquad (11)$$
$$Br + O \rightarrow Br + O_2 \qquad (12)$$

The bromine atoms are generated in a very similar manner to the chlorine atoms except that photolysis is also a major dissociative process for CH_3Br

$$CF_3Br + h\nu \rightarrow CF_3 + Br \qquad (13)$$
$$CH_3Br + h\nu \rightarrow CH_3 + Br \qquad (14)$$

The lifetimes of these halogen source gases cover a very large range from a few days to well in excess of 100 years. Generally speaking, the naturally emitted compounds, which contain hydrogen atoms, are susceptible to reactions in the troposphere with hydroxyl radicals and have short lifetimes, typically less than two years, whereas the anthropogenic compounds (eg CFC 11, CFC 12, CFC 113) have longer lifetimes because they are fully halogenated. The chlorofluorocarbons are all long-lived because they are mostly destroyed by photochemical processes in the stratosphere at altitudes where there is enough ultra-violet light. Atmospheric motion transports the CFCs in the troposphere up to the region where they can be photolysed, but they spend a relatively small proportion of the time

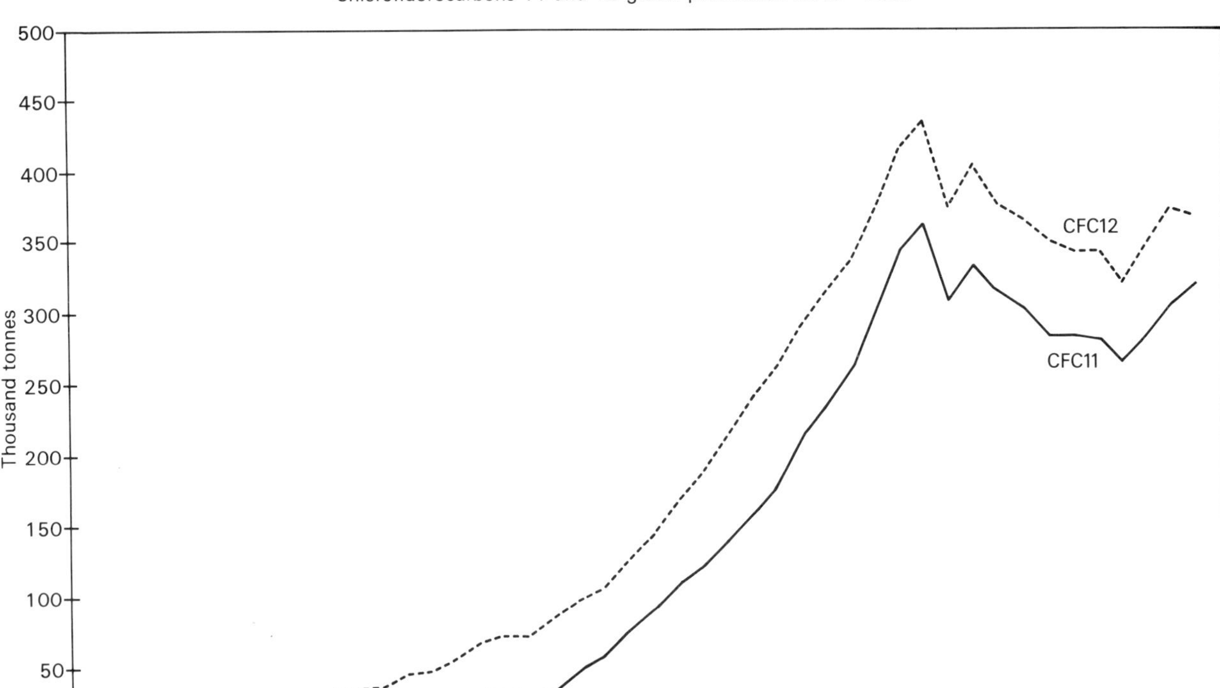

Figure 3.2 Annual production of CFC11 and CFC12, from the Chemical Manufacturers Association. Other producers (for example, Eastern Bloc countries) are not included

there before being transported down again. It takes a long time for transport to cycle the tropospheric content through to the appropriate altitude region.

The destruction of the various CFCs is strongly altitude dependent; this leads to somewhat different lifetimes (depending on the time for transport to the critical altitude) for the different compounds, eg 65 years for CFC 11, and 111 years for CFC 12 which is photodissociated more slowly at a given altitude. The long life of these compounds and their continued extensive emissions means that their mixing ratios in the atmosphere are increasing at steady rates (Figure 3.1, Table 3.1). The largest percentage increases are occurring for the molecules CHF_2Cl, $CF_2Cl.CFCl_2$ and CH_3CCl_3, (11.7%, 10% and 8.7% per year respectively), but the largest increase in abundance is occurring for CF_2Cl_2 at 17.1 pptv/year (1 pptv = 1 part in 10^{12} by volume). Production of this compound and $CFCl_3$ declined between 1974 and 1982 but has recently shown an increasing trend (Figure 3.2). The time to double the present atmospheric burden of these latter two compounds is approximately 20 years, assuming that emissions remain constant.

Evidence for the chemical decay of the chlorine and bromine containing molecules in the atmosphere has been collected in experiments which compare their mixing ratios in the northern and southern hemispheres and in the troposphere and the stratosphere. Figure 3.3 shows that $CFCl_3$, which has no removal mechanism in the troposphere except transfer into the ocean, is relatively evenly distributed between the northern and southern hemisphere. The small difference reflects the fact that emission takes place mostly in the northern hemisphere and the time for transfer of air

Table 3.1 Trends in surface mixing ratio of selected halocarbons

Molecule	Year	Average mixing ratio pptv	Increase %/year	Increase pptv/yr	Global production M kg	Year
CF_2Cl_2	1983	320	6	17.1	444	1982
$CFCl_3$	1983	200	5.7	9.6	310	1982
CHF_2Cl	1980	52	11.7	10	206	1984
$CF_2ClCFCl_2$	1985	32	10		138	1984
CH_3CCl_3	1983	120	8.7	8.6	545	1983
CCl_4	1979	140	1.8	2.1	830	1983

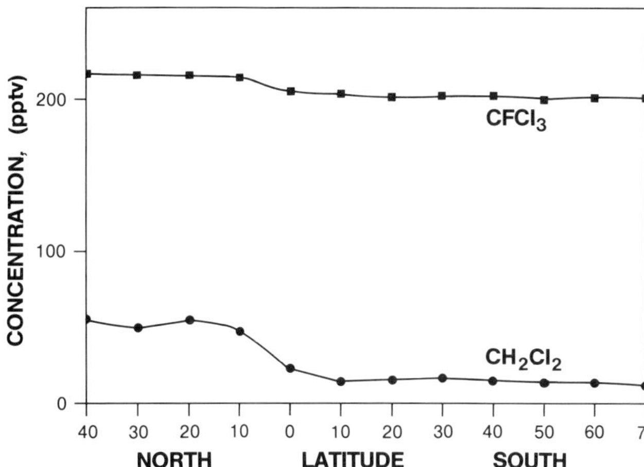

Figure 3.3 Latitudinal variation of the concentration of $CFCl_3$ and CH_2Cl_2 over the Atlantic Ocean in 1984 (data from Penkett and Rycroft, unpublished)

between the two hemispheres is slightly larger than one year. Figure 3.3 also shows that there is substantially more CH_2Cl_2 in the northern hemisphere than the southern hemisphere. This is caused by removal of much of the anthropogenic emissions by reaction with hydroxyl radicals during the time taken for inter-hemispheric transfer.

Figure 3.4 shows that $CFCl_3$ is removed by chemical processes in the stratosphere including photolysis and reaction with excited oxygen atoms, $O(^1D)$. Photolysis is the main removal mechanism and $CFCl_3$ has fallen to 0.1% of its tropospheric mixing ratio at an altitude of 29 km. Its overall atmospheric lifetime by stratospheric chemical removal is approximately 65 years. Many chlorine containing molecules are destroyed in this manner, releasing chlorine atoms which can take part in the chlorine ozone removal cycle shown above. This process has led to a significant increase in the mixing ratio of the most abundant form of inorganic chlorine, namely HCl, in the stratosphere in recent years.

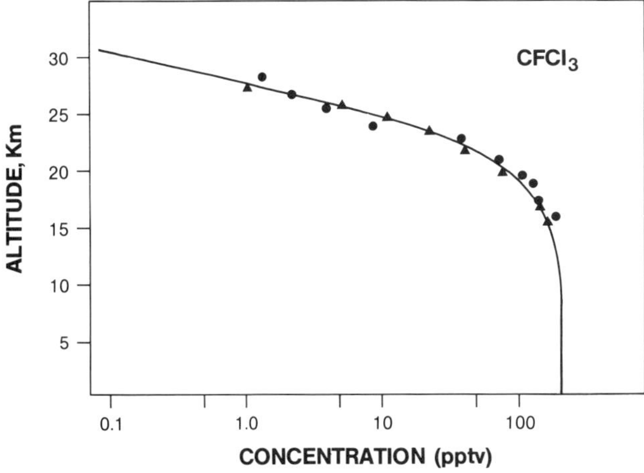

Figure 3.4 Vertical profiles of $CFCl_3$ obtained at 44°N in October 1982 (▲) and September 1983 (●). The solid line represents the bast fit to the data. (Schmidt, Knapska, Penkett and Hough, unpublished)

NITROUS OXIDE

Nitrous oxide is the predominant source gas for the reactive oxides of nitrogen, NO and NO_2, which are intimately involved in the NO-ozone removal cycle in the stratosphere:

$$N_2O + O(^1D) \rightarrow NO + NO \qquad (15)$$
$$NO + O_3 \rightarrow NO_2 + O_2 \qquad (16)$$
$$NO_2 + O \rightarrow NO + O_2 \qquad (17)$$

The N_2O mixing ratio in the troposphere was 304 ppbv in 1984 and this appears to be increasing at a rate of 0.7 ppbv per year, due mainly to combustion of coal and oil. The additional source is quite small in terms of the atmospheric burden but since the lifetime of N_2O is long ($\sim$ 150 years), it will continue to accumulate for many years before equilibrium is reached. Currently the combustion source is estimated to be 30% of the total. The natural sources are now thought to be mostly land based in the tropical regions of the earth, with a minor contribution from the oceans, probably concentrated in areas of ocean upwelling. The original concern over the growth of N_2O due to extensive use of fertilizers now appears to be much diminished.

METHANE AND CARBON MONOXIDE

Both methane and carbon monoxide can have considerable effects on ozone. Methane is a source of water in the stratosphere on oxidation and thus can contribute to HO_x production and thereby ozone removal:

$$CH_4 + OH \rightarrow CH_3 + H_2O \qquad (18)$$
$$H_2O + O(^1D) \rightarrow OH + OH \qquad (19)$$
$$OH + O_3 \rightarrow HO_2 + O_2 \qquad (20)$$
$$HO_2 + O \rightarrow OH + O_2 \qquad (21)$$

It also produces hydroxyl radicals (OH) directly by reaction with $O(^1D)$

$$CH_4 + O(^1D) \rightarrow OH + CH_3 \qquad (22)$$

and HCl, which is the main chlorine reservoir in the stratosphere, by reaction with chlorine atoms:

$$CH_4 + Cl \rightarrow HCl + CH_3 \qquad (23)$$

Hydroxyl radicals are responsible for removing many source gases in the troposphere. Methane and carbon monoxide are the major sink molecules for hydroxyl there (see Chapter 4, Figure 4.4) and thus they can indirectly affect stratospheric ozone by altering the lifetime of these source gases. This is particularly true of CH_3Cl and CH_3CCl_3, most of which are removed by reaction with OH in the troposphere. Another effect of CH_4 and CO is to produce ozone in the troposphere during oxidation in the presence of NO_2. (See Figure 4.4.)

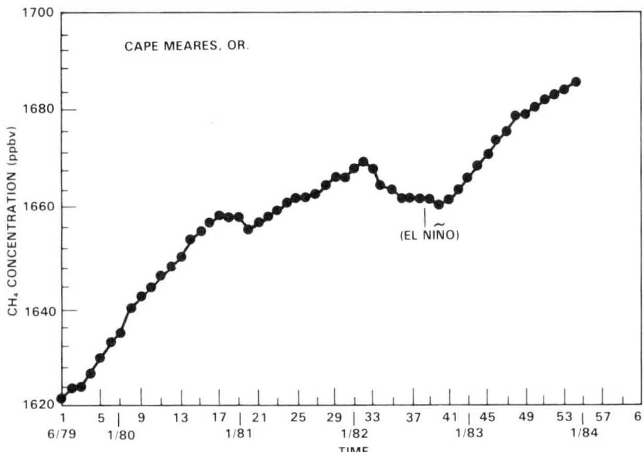

Figure 3.5 Trend in methane concentration measured at the ALE/GAGE station at Cape Meares, Oregon. The seasonal variation has been removed

Careful measurements made of methane in the atmosphere since the late 1970s (Figure 3.5) have shown that this gas is increasing at a steady rate of about 1.0% per year. Very recently a re-analysis of old spectroscopic data suggests that the mean tropospheric mixing ratio in the northern hemisphere was 1.14 ppmv in 1951, compared to present day values in the range of 1.6 ppmv. According to measurements of methane in air stored in ice cores, this increase has been occurring for over one hundred years and unperturbed levels of atmospheric methane were in the region of 0.7 ppmv (Figure 3.6).

Methane has both natural and anthropogenic sources. The increase in mixing ratio is the result of increased production from intensive agriculture (particularly of rice) and increasing extraction of fossil hydrocarbons. The exact breakdown of source type is in some dispute, but it is hoped that measurements of the C^{13}/C^{12} ratio in atmospheric samples will ultimately resolve the matter. At present these suggest that combustion and fossil fuel sources are unexpectedly important; ie emissions from natural gas usage and well-head operations are making a large contribution to the increase in atmospheric methane. An alternative explanation of the methane observations is that the concentration of hydroxyl radicals, which constitute the main chemical sink, has been lowered globally. Unravelling the correct interpretation of the methane growth will be difficult because of difficulties in quantifying all the various processes.

It is also suspected that the mixing ratio of carbon monoxide is increasing in the atmosphere, although the data are much less accurate than those for methane. Carbon monoxide has both natural and anthropogenic sources which include fossil fuel combustion, combustion of vegetation of various sorts, and oxidation of methane in the atmosphere. Oxidation of natural hydrocarbon appears to be the major natural source of CO, but as yet this remains a very uncertain quantity because of the scarcity of data on the concentration of these compounds in the atmosphere.

CARBON DIOXIDE

Carbon dioxide is chemically unreactive in both the troposphere and the stratosphere. However it is radiatively active, and cools the stratosphere whilst warming the troposphere. Its main effect on stratospheric ozone is therefore due to its influence on stratospheric temperature, and thus the rates of the many reactions involved in ozone chemistry. Carbon dioxide was the first trace gas for which observations showed an increasing trend in atmospheric abundance. The classic work of Keeling at Mauna Loa Laboratory in Hawaii produced the data set shown in Figure 3.7. The cause of the upward trend is undoubtedly the continuous emission from the combustion of fossil fuel. However the annual variation, which is much more pronounced in the northern hemisphere, could have a number of different causes and the precise interpretation is not clear at present. The most likely cause is the seasonal cycle in biological activity at mid-latitudes in the northern hemisphere. This has the effect of reducing the atmospheric CO_2 concentration during the summer months. Another possibility is that increased emissions of CO_2 from the combustion of fossil carbon leads to an increase in the average CO_2 concentration during the winter. Careful examination of the data

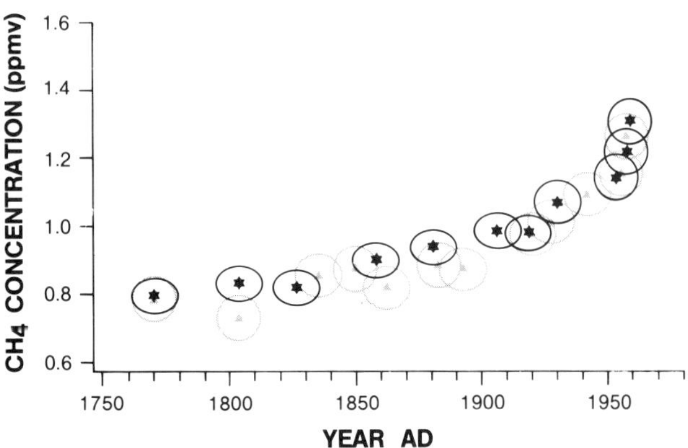

Figure 3.6 Historic record of methane concentration in air stored in ice cores from Antarctica. Data from Stauffer et al, 1985

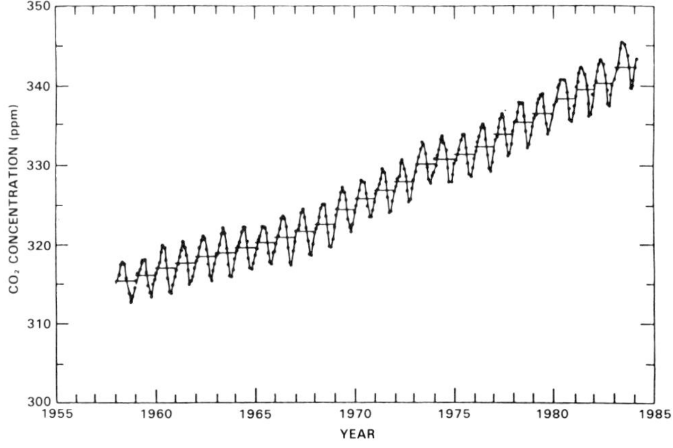

Figure 3.7 Concentration of atmospheric CO_2 at Mauna Loa, Hawaii. (Keeling, Scripps Institute of Oceanography, USA)

record shows that the amplitude of the seasonal variation has been increasing in recent years, at several northern hemisphere locations. This has been interpreted specifically in terms of enhanced activity of the biosphere over a period when the climate is warming.

NON-METHANE HYDROCARBONS AND NITROGEN OXIDES IN THE TROPOSPHERE

Most of the non-methane hydrocarbons and nitrogen oxides in the troposphere are unlikely to find their way into the stratosphere, since their lifetimes are too short. It is possible, however, that they can increase the amount of ozone present in the troposphere, particularly in the northern hemisphere, by the same mechanism which produces photochemical smog. The chemical mechanism is similar to that shown in Figure 4.4 but for higher hydrocarbons. The latitudinal distribution of ethane is similar to CH_2Cl_2 (Figure 3.3) with larger mixing ratios being observed in the northern hemisphere. More-reactive hydrocarbons with more carbon atoms show a much more pronounced interhemispheric concentration gradient. Also large seasonal dependencies in mixing ratios of many hydrocarbons have been observed in the northern hemisphere with higher values being observed in the winter.

There are very limited data on NO and NO_2 in the more remote parts of the troposphere. It has been shown though that NO concentrations are less than 5 pptv over the eastern equatorial Pacific Ocean. Similarly the NO_2 concentration measured at an altitude of 3 km on Hawaii in June showed an average concentration of 30 pptv. In mid-latitudes average mixing ratios of about 100 pptv were observed at a coastal site in Ireland in June and 280 pptv and 300 pptv respectively were observed in winter and summer at Niwot Ridge, Colorado. These latter three mixing ratios would provide sufficient NO_x for substantial ozone formation to occur during the course of photochemical destruction of the hydrocarbons observed in the troposphere.

CHEMISTRY 4

Models of stratospheric chemistry require accurate knowledge of the reaction mechanisms and the rate coefficients for all reactions involved in the chemistry of ozone and associated trace gases. A much improved data base and quantitative understanding of the processes involved has resulted from studies carried out over the past decade. The basic reactions leading to ozone production and loss, and the reactions of the source gases, have been introduced in Chapters 2 and 3. Figure 4.1 to 4.4 show schematic representation of our current knowledge of the atmospheric reactions involving HO_2, NO_X, ClO_X and methane oxidation.

REACTIONS AND RATE COEFFICIENTS

All of the reactions in the catalytic cycles for O_3 destruction, described in Chapter 2 which involve reactions (5) and (6) for X = H, OH, NO and Cl, have now been thoroughly investigated in the laboratory. The data from the different studies have been generally in good agreement and consequently the assigned uncertainties in the values of the rate coefficients at 298 K are all less than ±30%, the largest uncertainty being associated with the reaction of OH with O_3 (NASA/WMO, 1986). It should be recognised however that the calculated upper stratospheric ozone profile is particularly sensitive to the rate coefficients of the HO_X reactions. Also the errors in the rate coefficients increase as the temperature diverges from room temperature and for many reactions measurements do not exist for temperatures less than 220 K. The increased uncertainty at lower temperatures is illustrated in Table 4.1 which shows uncertainties calculated from the standard deviations on the rate coefficients given in the NASA evaluation, at 298 K and 220 K (appropriate for the mid-stratosphere). Data for X = Br are also included; the large uncertainty on the O + BrO reaction arises because only one measurement of this rate coefficient has been reported.

Table 4.1 Uncertainty factors in the rate coefficients for key reactions involved in the stratospheric ozone destruction

X rate coefficient for ($X + O_3$)			XO rate coefficient for (XO + O)		
	298 K	220 K		298 K	220 K
H	1.25	1.59	OH	1.2	1.35
OH	1.3	1.86	HO_2	1.2	1.52
NO	1.15	1.30	NO_2	1.3	1.46
ClO	1.2	1.52	ClO	1.1	1.31
Br	1.2	1.52	BrO	3.0	4.04

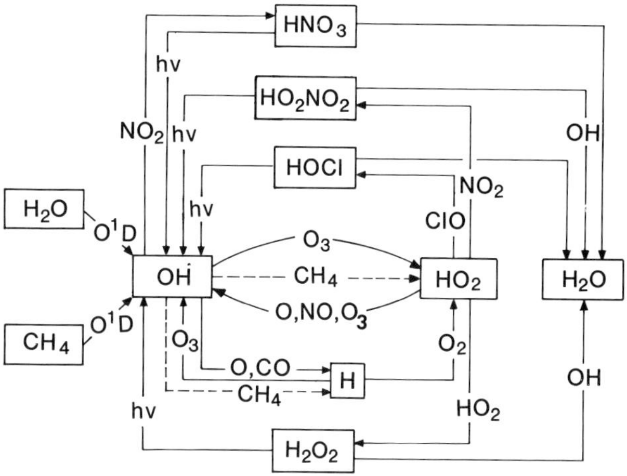

Figure 4.1 Schematic of odd-hydrogen chemistry

Ozone destruction also occurs in cycles not involving participation of atomic oxygen eg:

$$OH + O_3 \rightarrow HO_2 + O_2 \quad (20)$$
$$HO_2 + O_3 \rightarrow OH + 2O_2 \quad (24)$$

$$\text{net: } 2O_3 \rightarrow 3O_2$$

$$NO + O_3 \rightarrow NO_2 + O_2 \quad (16)$$
$$NO_2 + O_3 \rightarrow NO_3 + O_2 \quad (25)$$
$$NO_3 + h\nu \rightarrow NO + O_2 \quad (26)$$

$$\text{net: } 2O_3 \rightarrow 3O_2$$

These cycles are more important in the lower stratosphere and, ahtough all the reactions have been studied, there is still considerable uncertainty in the rates of reactions (24) and (26).

There have been a number of changes in rate coefficients recommended for several important reactions, particularly involving reservoir species controlling the abundance of radicals, which have led to changes in predictions of ozone depletion. These are considered in the following section, together with some novel chemical aspects which have been discussed very recently in connection with ozone behaviour in the Antarctic stratosphere (see Chapter 10).

SINKS AND RESERVOIRS

The removal of radicals from the stratosphere occurs by chemical conversion to more stable oxidation products. For example, NO_2 is converted to nitric acid by reaction with OH radicals

$$OH + NO_2 + M \rightarrow HNO_3 + M \quad (27)$$

Subsequently HNO_3 has two fates. It can react in the stratosphere to reform active nitrogen (HNO_3 acting as a reservoir) or it can be transported down into the tropo-

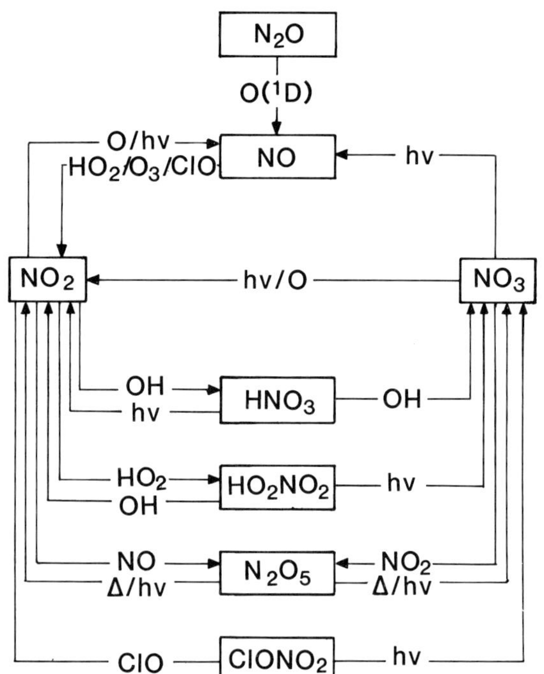

Figure 4.2 Schematic of odd-nitrogen chemistry

sphere where it is removed in rain (HNO_3 acting as a sink). HO_x, ClO_x and BrO_x radicals are removed in a similar way in the form of reservoir species such as H_2O, HCl and HBr respectively.

An example of the key role played by reactions of reservoir species is the reaction:

$$OH + HCl \rightarrow Cl + H_2O \quad (28)$$

This serves to convert chlorine tied up as stable HCl, to active ClO_x which can participate in catalytic cycles leading to removal of ozone. The rate coefficient has been measured numerous times by straightforward direct techniques. However, two recent careful measurements have led to an upward revision of its value at 298 K by 25%. This has a significant effect on the calculations of ozone depletion.

Another important revision of chemical data in recent years involves the reactions of hydroxyl radicals with nitric acid and peroxynitric acid:

$$OH + HNO_3 \rightarrow H_2O + NO_3 \quad (29)$$
$$OH + HNO_4 \rightarrow H_2O + NO_2 + O_2 \quad (30)$$

As well as releasing active NO_x from the reservoirs, these reactions convert OH to H_2O and are important loss processes for active HO_x in the lower stratosphere. Changes in the recommended rate constants for reactions (29) and (30) have led to significant revision of the calculated depletion of the ozone column in recent years.

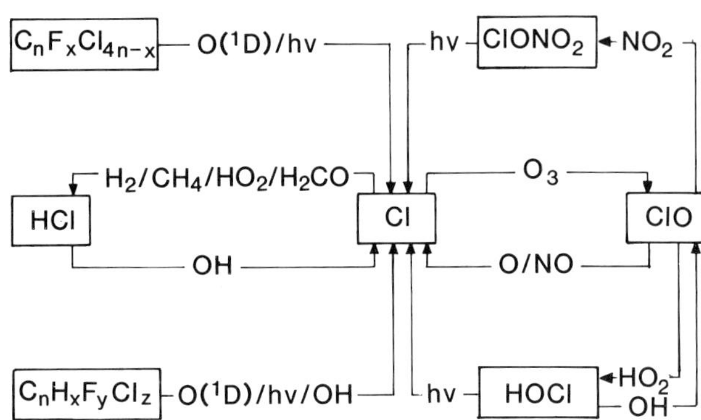

Figure 4.3 Schematic of odd-chlorine chemistry

For loss of ClO_x and NO_x the key reactions are

$$Cl + CH_4 \rightarrow HCl + CH_3 \quad (23)$$
$$OH + NO_2 + M \rightarrow HNO_3 + M \quad (27)$$

These and other reactions of ClO_x and NO_x radicals with H-containing molecules are now generally well understood and only minor changes in the recommended rate coefficients have occurred in recent years.

TEMPORARY RESERVOIRS

The chemistry in the stratosphere, particularly at lower altitudes, is complicated by the involvement of less stable temporary reservoir species such as HOCl, H_2O_2, HNO_4, N_2O_5 and $ClONO_2$ which store active radicals. The molecules result in strong coupling between the different families of gases, so that the efficiency with which a particular catalytic cycle acts depends on the abundance of

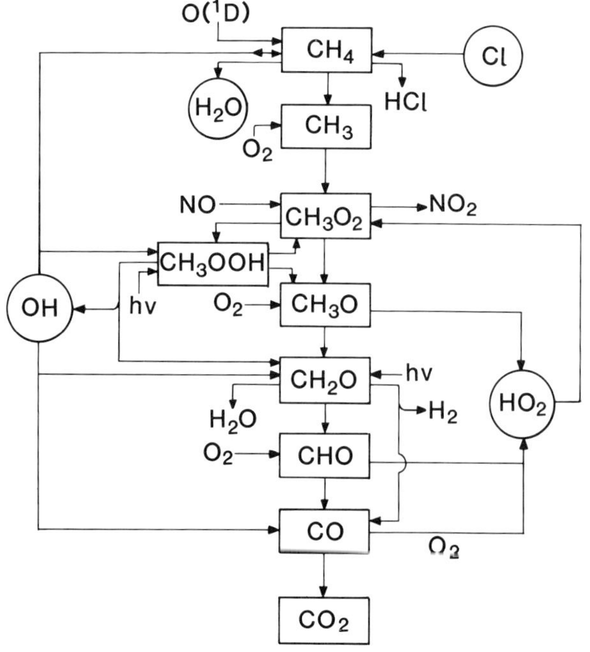

Figure 4.4 Schematic of atmospheric methane oxidation chemistry

other trace species. This also occurs through such reactions as:

$$NO + ClO \rightarrow NO_2 + Cl \qquad (31)$$
$$HO_2 + NO \rightarrow NO_2 + OH \qquad (32)$$

These serve to interconvert active NO_x, HO_x and ClO_x species without participation of O or O_3 and therefore reduce the efficiency of the catalytic cycles. Because of their importance these reactions have been subjected to careful study and are now well understood. Coupling between families also occurs through the formation of temporary reservoir species, specifically chlorine nitrate ($ClONO_2$), hypochlorous acid (HOCl), peroxynitric acid (HO_2NO_2) and nitrogen pentoxide (N_2O_5). The presence of all these species has been postulated on the basis of laboratory studies but recent atmospheric measurements provide strong evidence that $ClONO_2$ and N_2O_5 are indeed present in the stratosphere. The main chemical processes involved in the formation of the temporary reservoir species can be seen from the schematic diagrams, Figures 4.1 to 4.4. A typical example which has attracted much discussion is the formation of $ClONO_2$:

$$ClO + NO_2 + M \rightarrow ClONO_2 + M \qquad (33)$$

This serves to tie up both active NO_x and ClO_x in the mid-stratosphere and has a key role in determining their catalytic activity at these altitudes. An issue of considerable debate has been the possibility of the formation of more than one isomer of $ClONO_3$ in reaction (33). Careful laboratory work has now confirmed that $ClONO_2$ formation is the only significant pathway.

Another temporary reservoir species postulated to be important in the lower stratosphere is hydrogen peroxide, H_2O_2, which is formed by combination of HO_2:

$$HO_2 + HO_2 \rightarrow H_2O_2 + O_2 \qquad (34)$$

Although there have been a number of detailed studies of reaction 34, the rate and product channels under stratosphere conditions have not been fully characterised. There is now evidence for the presence of H_2O_2 in the stratosphere, but the importance of H_2O_2 chemistry remains uncertain.

PHOTOCHEMICAL REACTIONS

It will be seen from the schematic illustrations, Figures 4.1 to 4.4, that photochemical dissociation processes are important pathways for break-down of temporary reservoirs as well as source gases. The calculation of the rates of these processes requires knowledge of the absorption cross sections and the quantum yields for all product channels in addition to the wavelength dependent solar flux. The absorption cross sections are almost certainly temperature dependent but the majority of the absorption cross section measurements have been carried out at 298 K. Therefore, it is necessary to measure absorption cross sections at stratospheric temperatures particularly for molecules such as HNO_3, $ClONO_2$ and HO_2NO_2. It is worth noting that these molecules are hard to manipulate in the laboratory. Generally measurements of the photochemical rate parameters have not received attention commensurate with their importance.

Release of active chlorine from the halocarbon source gases in the stratosphere mainly occurs by direct photolysis. Release of NO from N_2O and OH from H_2O requires the participation of $O(^1D)$ produced from ozone photodissociation:

$$O_3 + h\nu\ (\lambda < 320\ \text{mm}) \rightarrow O(^1D) + O_2 \qquad (3)$$

$O(^1D)$ also reacts with N_2O to give $N_2 + O_2$ and is quenched to ground state O atoms by collision with the major atmospheric gases N_2, O_2 and CO_2. Thus several elementary steps are involved in calculation of the overall production of HO_x and NO_x radical species from the source gas. Although the rates of the elementary processes are well determined, the cumulative errors in successive steps tend to increase the uncertainty for the overall process to an extent that interpretation of atmospheric measurements is made difficult. The prospects for further reduction of the errors of many of the elementary steps are limited, and the possibility of reducing this uncertainty using experiments simulating the overall process is currently being considered.

BROMINE CHEMISTRY

Although stratospheric Br and BrO destroy ozone in an analogous manner the ClO_x species, bromine chemistry differs from chlorine chemistry in several important aspects. Because the H-Br bond strength is much less than the H-Cl bond strength, hydrogen abstractions to produce HX tend to be much more rapid for chlorine than for bromine. Indeed, reactions such as $X + CH_4$ and $X + H_2$ which are important for X = Cl are endothermic for X = Br and can be neglected. As a result BrO is expected to be the dominant form of BrO_x in the stratosphere and the BrO and BrO + ClO reactions take on particular importance in the lower stratosphere despite the relatively low BrO_x mixing ratio. The key steps are:

$$BrO + BrO \rightarrow 2Br + O_2 \qquad (35)$$
$$BrO + ClO \rightarrow Br + Cl + O_2 \qquad (36)$$

which when combined with:

$$3(Br + O_3 \rightarrow BrO + O_2) \qquad (11)$$
$$\text{and } Cl + O_3 \rightarrow ClO + O_2 \qquad (9)$$

yields net: $4O_3 \rightarrow 6O_2$

Thus ozone can be removed via a chain mechanism which does not require oxygen atoms, unlike the cycles in reactions (6) and (7). This mechanism can work therefore in the lower stratosphere. There remain considerable uncertainties in the reaction rates and pathways for BrO_x reactions compared to those of ClO_x. Of particular importance is the temperature dependence and the alternate product channels in the BrO + BrO and BrO + ClO reactions, particularly in view of the Antarctic ozone reductions discussed in Chapter 10.

ORGANIC MOLECULES

The HO_x budget in the atmosphere is closely linked with and influenced by the oxidation of organic compounds. The major organic constituent in the atmosphere is methane, CH_4, and its dominant sink is its reaction with OH ultimately producing water vapour, molecular hydrogen and CO_2. As noted above this is a major source of stratospheric water vapour, although most of the methane is in fact oxidised in the troposphere. The detailed chemistry of methane oxidation is now reasonably well understood. The reactions involved are illustrated in Figure 4.4. For the stratosphere below 35 km the oxidation scheme is simplified by the presence of sufficient NO for complete conversion of the peroxyradicals by reaction with NO. This makes CH_4 oxidation a net source of ozone through the reactions:

$$CH_3O_2 + NO \rightarrow CH_3O + NO_2 \quad (37)$$
$$NO_2 + h\nu \rightarrow O + NO \quad (38)$$
$$O + O_2 + M \rightarrow O_3 + M \quad (2)$$

This oxidation process is also a source of HO_x through the photolysis of formaldehyde formed by oxidation of CH_3O:

$$CH_3O + O_2 \rightarrow HO_2 + HCHO \quad (39)$$
$$HCHO + h\nu \rightarrow H + HCO \quad (40)$$
$$H_2 + CO \quad (41)$$

The rate coefficients for these reactions are all reasonably well established. In the troposphere, where NO concentrations can be very low, other reactions of CH_3O_2 become important for example:

$$CH_3O_2 + HO_2 \rightarrow CH_3OOH + O_2 \quad (42)$$

The kinetics and products of this reaction are not well defined and further work is needed on this key tropospheric reaction. Uncertainties in reaction rates and mechanisms exist for other hydrocarbons such as C_2H_6 (ethane), C_3H_8 (propane) and C_2H_2 (acetylene), which are useful as tracers to test the transport and chemistry used in models. Ethane and propane oxidation leads to the formation of peroxyacetylnitrate (PAN), which is a reservoir for reactive nitrogen in the troposphere and lower stratosphere.

Halocarbons are oxidised in the stratosphere in a sequence of radical reactions initiated by the photolysis of the molecule, by its reaction with $O(^1D)$ or OH. The kinetics of the elementary reactions involved in the oxidation mechanism were not known until recently and it has been generally assumed in model calculations that all the chlorine atoms of the CFC molecule are released simultaneously in the atmosphere with a negligible delay following the initial photolysis or $O(^1D)$ attack. The new data provide further information on the reaction mechanisms for CFC, but for CH_3Cl and CH_3CCl_3 the mechanism is not well known. Preliminary calculations indicate that the assumption of simultaneous Cl atom release is only likely to lead to the overestimate of a few percent in the stratospheric Cl_x concentration, and consequently a negligible effect on the calculation of ozone depletion.

Halogenated hydrocarbons such as CH_3Cl, CH_3CCl_3 etc can be oxidised in the troposphere by reaction with OH. However, this reaction is relatively slow for most of these compounds and consequently their input in the stratosphere is still significant.

HETEROGENEOUS REACTIONS

Heterogeneous reactions of the gases involved in stratospheric chemistry often occur in laboratory studies due to the reactive surfaces of vessels. In the atmosphere where the surface is provided by airborne particulate and the surface-areaa/volume ratio is much lower, the rates of these reactions are substantially reduced. The present knowledge of these processes is, however, seriously inadequate, so that a reliable assessment of the impact on stratospheric ozone of heterogeneous reactions of trace species on the surface of aerosol particles cannot be made. Such heterogeneous reactions are most likely to be significant for the slow reactions involving temporary reservoir species, particularly in their reactions with water, which is present in the aerosol. Of particular interest are reactions of chlorine nitrate with HCl and with water:

$$ClONO_2 + HCl \rightarrow Cl_2 + HNO_3 \quad (43)$$
$$ClONO_2 + H_2O \rightarrow HOCl + HNO_3 \quad (44)$$

The reactions serve to release active ClO_x and HO_x from the stable reservoir species HCl and H_2O, through photoylsis of the reaction products.

Assessment based on the limited amount of currently available information on these reactions indicates that they are not important for the global average atmospheric situation. Local effects resulting from enhanced aerosol concentration due to volcanic injections or polar stratospheric clouds could be considerably more significant (see Chapter 10). For example polar night clouds may influence the chemistry of the chlorine and nitrogen-containing temporary reservoir species at high latitudes through the heterogeneous reactions of N_2O_5 and chlorine nitrate.

Further characterisation of these heterogeneous reactions in the laboratory may modify our formulation of the behaviour of temporary reservoir species in the future.

SUMMARY

There has been a continued steady improvement in the laboratory data for the reaction rate coefficients, product distributions, absorption cross sections and photodissociation quantum yields of the elementary chemical processes controlling stratospheric ozone. Changes in the accepted rate coefficients for several important reactions eg O + ClO, OH + HCl and OH + HNO_3 have led to significant changes (in both directions) in the predicted ozone depletions due to chlorine and nitrogen oxides in the stratosphere.

RADIATION, DYNAMICS AND TRANSPORT 5

RADIATION

In the absence of motions, the thermal structure of the atmosphere is determined by a balance between absorption of solar radiation and cooling by terrestrial emission in the infra-red. Large departures from this simple balance are induced by dynamics. Radiative processes must be treated properly if temperatures are to be reproduced and atmospheric waves properly damped in numerical models. Moreover, the rates of many chemical reactions depend strongly on temperature.

Figure 5.1 shows the variation of temperature with height in the atmosphere. Boundaries between different atmospheric regions are marked by sharp changes in the temperature variation (or lapse rate) with height. The stratosphere is heated from above mainly by the absorption by ozone of solar ultra-violet radiation. The troposphere is heated mainly from below by thermal radiation emitted from the Earth's surface. Convection is triggered in the troposphere because warm air in contact with the ground underlies colder air aloft. In the stratosphere, on the other hand, convection is suppressed so chemicals reside there longer than they do in the troposphere.

The sun emits approximately the electromagnetic spectrum of radiation characteristic of a black-body at a temperature of 5800 K, while the Earth emits at a mean atmospheric temperature of about 245 K. The peak radiative fluxes from these two sources thus occur at well-separated wavelengths in the visible (400-600 nm) and infra-red (10,000-15,000 nm or 10-15 μm). Solar and terrestrial fluxes may thus be treated separately in calculating the heat balance. Figure 5.2 shows the relative importance of different atmospheric constituents as attenuators of radiation. It can be seen that radiation from the sun of wavelengths shorter than 290 nm does not get through the stratosphere, and that relatively transparent "windows" exist in both the visible and infra-red parts of the spectrum.

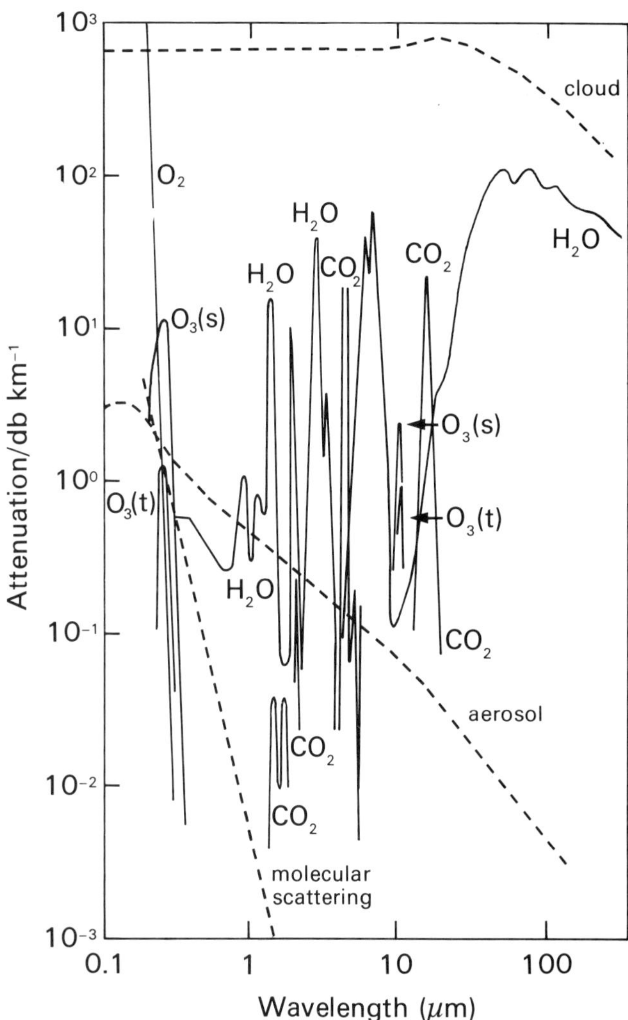

Figure 5.2 The relative importance of atmospheric constituents as attenuators of radiation. The attenuation by a 1 km horizontal path is shown on the ordinate

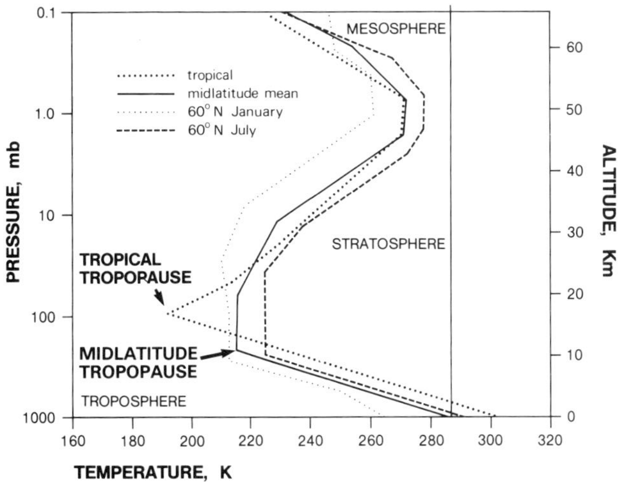

Figure 5.1 The vertical profile of temperature, and its use to define the troposphere and the stratosphere at different latitudes

The figure also shows that the molecules which affect the transmission in the infra-red are not the major constituents of the atmosphere, N_2 and O_2, but the much less abundant molecules CO_2, O_3 and (in the troposphere) H_2O. (The reason is to do with the type of transition between discrete energy levels that the molecules can make.) Liquid water and ice in the form of clouds are also radiatively important in both the visible and infra-red (dashed line on Figure 5.2). The contributions each gas makes to the balance of heating and cooling is shown as a function of altitude in Figure 5.3. The essential point is that without water vapour and carbon dioxide, infra-red radiation emitted from the surface would escape more readily than it actually does; the surface temperature is thus about 30°C higher than it would be for an atmosphere containing only N_2 and O_2. Ozone participates in this infra-red blanketing by its absorption at 9600 nm. Unlike CO_2 the atmospheric abundance of ozone below the stratopause is significantly affected by photochemical reactions.

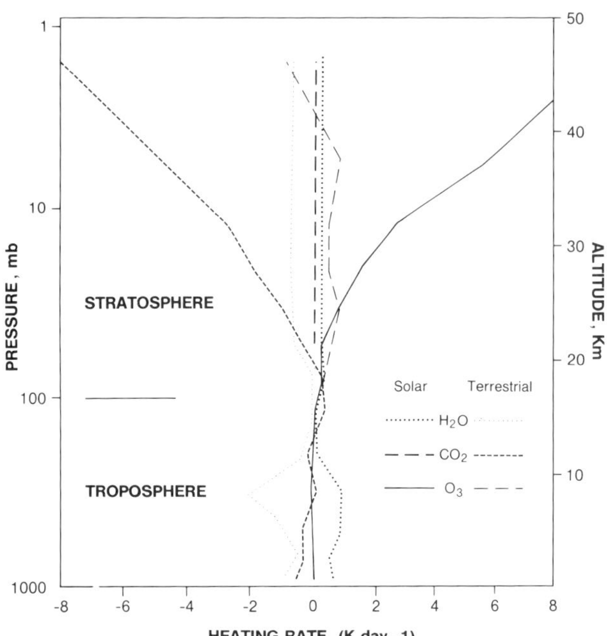

Figure 5.3 The contributions of water vapour, carbon dioxide and ozone to the vertical profile of heating and cooling in the tropics

Accurate calculations of solar heating rates in the atmosphere require knowledge of the incident radiation and absorption cross-sections of various gases. Knowledge of irradiance in the spectral region 175-320 nm has increased recently as a result of measurements by the SBUV instrument on the Nimbus 7 satellite, and by the SME satellite. Ozone cross-sections are known well enough for most purposes, but better measurements in the Huggins bands (around 360 nm) are needed. Probably the largest source of error in calculations of heating rates are uncertainties in measurements of the distribution and concentration of ozone. It is important to estimate likely errors since dynamical models require the net radiative heating, often a small difference between heating by solar radiation and cooling by infra-red radiation.

The major difficulties in calculating longwave radiation are that (1) gases can have a complicated spectrum of absorption; (2) spectroscopic data may not be accurate or complete; and (3) so-called band models (used for speed to circumvent calculations for individual lines of a complicated spectrum) may be unsatsifactory. Whereas simple models of absorption line shapes can duplicate atmospheric spectra at low pressures, agreement worsens as pressure increases. This poses a problem in remote sensing of the lower stratosphere. Greater uncertainties lie in the transmittances of a band of lines. Better measurements of these broadband transmittances are needed under stratospheric conditions. The radiative effects of trace gases such as N_2O and the chlorofluorocarbons (CFCs) require further study, as do the effects of aerosols and clouds in the lower stratosphere.

Studies of the radiative transfer are being advanced by the availability of satellite data on ozone, temperature, solar irradiance and terrestrial radiation. Accurate data obtained over a long time period are needed for a complete understanding of radiative processes in the middle atmosphere.

DYNAMICS AND TRANSPORT

There have been several recent advances in our understanding of dynamics and transport. Measurements from satellites have furnished global analyses of the large-scale circulation; ground-based radars and lidars have given information on gravity waves; general circulation models have had some success in reproducing the gross characteristics of the middle atmosphere; and the way transport should be represented in simplified numerical models has been clarified to some extent.

Global climatologies of the stratosphere and mesosphere (or "middle atmosphere") have been produced recently. Better data coverage in time and space has allowed more wave types in the atmosphere to be identified. A phenomenon that continues to attract attention is the stratospheric sudden warming, during which temperatures in the stratosphere may rise locally by more than 80 K in a few days. Besides disrupting the circumpolar westerly flow in winter stratospheric warmings are thought to produce significant poleward transport of trace gases. It is thought that warmings are initiated by large-scale disturbances in the troposphere, but their dynamics are not yet fully understood.

The dynamics of the middle atmosphere has frequently been interpreted using the theory of wave, mean-flow interaction. Planetary-scale waves, defined as departures from the zonal-mean flow, are taken to propagate on that flow, forcing changes in it. Such a separation into waves and a mean flow may not be appropriate, however, when the circulation is highly disturbed. Recently, a number of studies have been based instead on synoptic maps of potential vorticity, a fundamental quantity of dynamical meteorology. Potential vorticity does not change along an air parcel's trajectory for as long as friction and radiation can be neglected (for about a week in the stratosphere), nor does potential temperature. So contours of potential vorticity plotted on surfaces of constant potential temperature define material lines, allowing the movement of strings of air parcels to be followed. Such maps are therefore useful for studying how material is transported. Figure 5.4 gives an example. Contours of potential vorticity (material lines) are highly and irreversibly deformed as pools of air from low latitudes, having low values of potential vorticity, are drawn eastward and poleward around the westerly vortex.

Gravity waves are thought to have an important influence on the mesospheric circulation, affecting both the large-scale dynamics and transport in the region. The design of parametrization schemes to represent their effects in models is a major goal of current research.

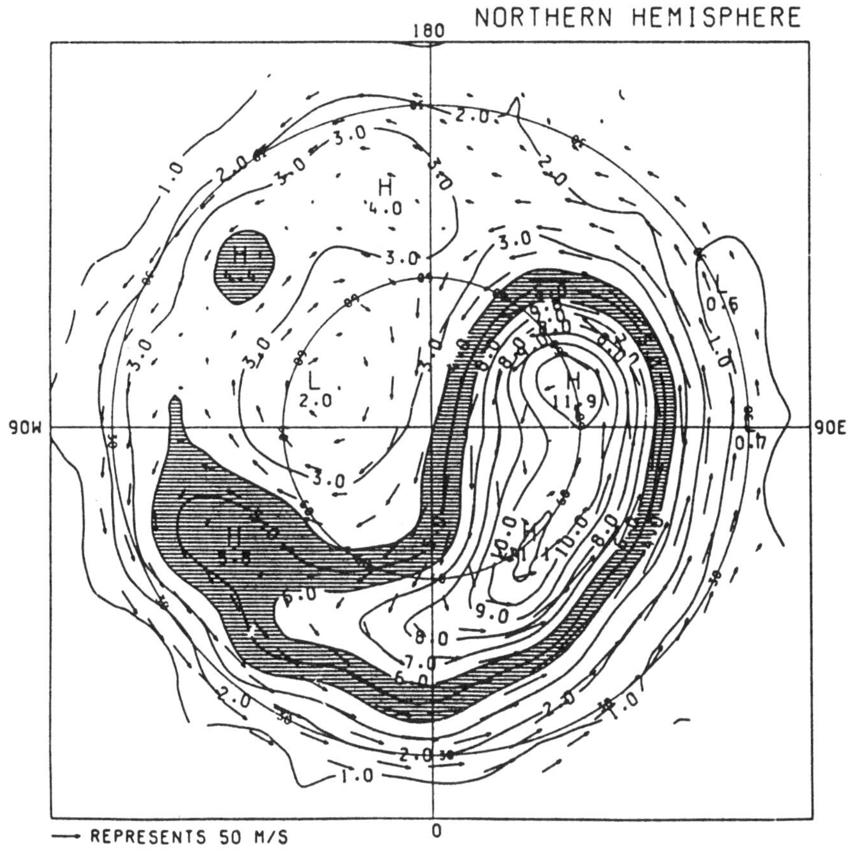

Figure 5.4 Potential vorticity and geostrophic winds on the 850K isentropic surface (near 10 mb, 30 km) for 4 December 1981

GENERAL CIRCULATION MODELS OF THE MIDDLE ATMOSPHERE

Many aspects of the circulation of the middle atmosphere have been reproduced by general circulation models (GCMs) – comprehensive three dimensional numerical models of the atmosphere, with the main physical processes represented in some way. They have been used in climatological and short-term (forecast) simulations, in studies of transport, and in numberical experiments in which a perturbation is applied to the system (for example, changing the concentration of carbon dioxide). Most GCMs still have deficiencies, however. The most serious is that, perhaps with one exception (a model with very high spatial resolution), they predict temperatures that are too cold in the middle atmosphere in winter, and westerly winds that are too strong. The problem is likely to be an inadequate treatment of dynamics, possibly the inability of the models to resolve the full range of gravity waves in the real atmosphere. Because of this difficulty, most GCMs currently in use have limited value in providing a parametrization of transport for simpler models.

THEORY OF TRANSPORT IN SIMPLIFIED MODELS

At present computer time is too expensive to include full photochemistry in GCMs. Much effort has therefore been directed to ways in which dynamical and transport processes might be parametrized in a two-dimensional model. Although we now understand better how this should be attempted, the physical basis for the method is still uncertain.

When considering transport in height and latitude, the meridional circulation relevant to transport is not defined by horizontal and vertical velocities averaged around a latitude circle. Instead, it is the so-called Brewer-Dobson circulation, the pole-to-pole circulation shown schematically in Figure 5.5. In the middle atmosphere this meridional circulation exists largely because of atmospheric disturbances. Parcels of air from widely dispersed origins (and therefore having different properties) may be found along a latitude circle, particularly at low latitudes. Air parcels appear "mixed" in a zonally-averaged view, an effect that is represented in simplified models by a parametrization based on an idealised mixing process.

The zonally-averaged transport in the troposphere, stratosphere and mesosphere may be summarised by reference to Figure 5.5. In the troposphere, there is advection by the Hadley circulation (the cell near the equator); quasi-horizontal dispersion by planetary and smaller scale eddies (thick horizontal arrows); and vertical mixing by convection (vertical arrows). Time scales for transport are mostly short (about a week to a few months). In the winter stratosphere, transport is largely dictated by planetary-scale eddies, both indirectly in a wave-driven meridional circula-

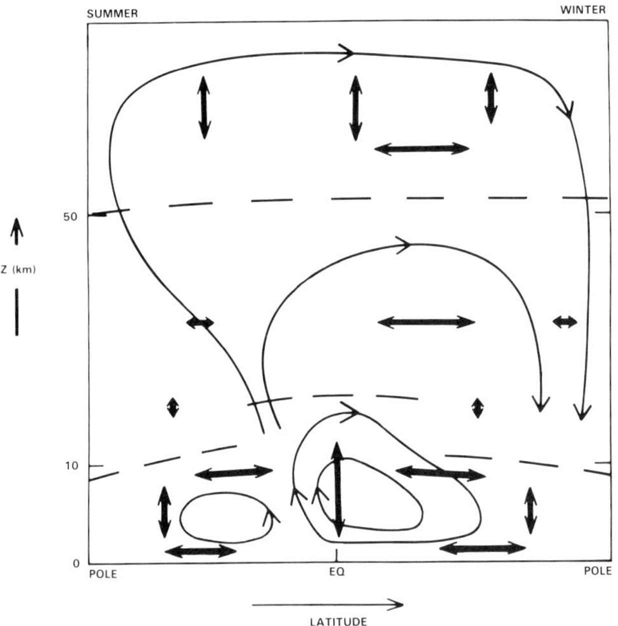

Figure 5.5 Schematic illustration of zonally-averaged transport up to the mesopause. Single arrows denote mean circulation; double arrows denote quasi-horizontal and vertical diffusion. Pecked lines mark the tropopause (lower) and the stratopause (upper)

tion, and directly by quasi-horizontal dispersion or "mixing". In the mesosphere, planetary-scale eddies decrease in amplitude and transport processes appear to be dominated by gravity waves. They reach large amplitudes as they propagate upwards from the troposphere (because density decreases with height) and they may "break". The turbulence, generated as light and heavy fluid elements overturn, mixes trace species in the vertical, and also leads to advection in a strong pole-to-pole circulation.

One of the main limitations of the parametrization of eddies in the way outlined above is that the simplified model cannot be interactive in the sense of predicting changes in circulation resulting from, say, a change in the concentration of a radiatively active gas.

STRATOSPHERE-TROPOSPHERE EXCHANGE

Photochemical studies of the middle atmosphere rely on sound knowledge of the amounts of chemicals that are imported into and exported from the region. Unfortunately, the fluxes of trace gases across the tropopause are not well known. Motions on many scales in the troposphere may involve the transfer of air between it and the stratosphere. The strongest ascent in the troposphere is driven by convection and is centred over the tropical Western Pacific. Virtually all the water vapour in the stratosphere is believed to enter it near here, passing through the coldest tropospheric temperatures that occur anywhere on the globe. It has been proposed that deep cumulonimbus convection, and radiatively driven circulations in large anvils atop thunderstorm clouds may be the principal mechanisms for exchange. More measurements from aircraft are needed to confirm this hypothesis.

Outflow of air from the stratosphere is believed to be mainly by the folding of the tropopause when cyclonic storms develop at extratropical latitudes. The tropopause steepens near the core of an upper level jet stream. Descending air on the poleward side of the jet is stratospheric, and is richer in ozone than ascending air on its equatorward side. It is also thought that stratospheric air enters the troposphere when air moves irreversibly along isentropic surfaces which intersect the tropopause, for example, in the region of "cut-off" lows.

SUMMARY

Radiation transport and photochemical reactions on many spatial and temporal scales affect the composition and distribution of trace gases in the atmosphere. The way these processes are inter-related is summarised in Figure 5.6. There are many inadequacies in our current understanding of them. Forecasting the future composition of the atmosphere in the style of numerical weather forecasting has not yet been done.

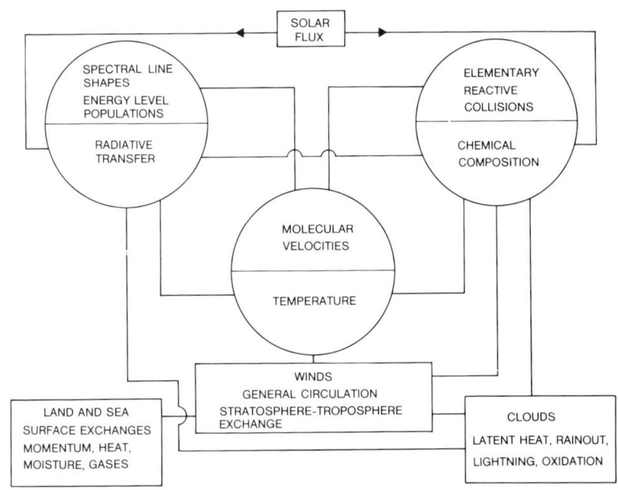

Figure 5.6 Schematic illustration of the relationship between dynamics, radiation and chemistry

MEASUREMENTS OF OZONE 6

INTRODUCTION

The integrated amount of ozone in a vertical column extending from the earth's surface to the top of the atmosphere is referred to as the total amount of ozone, total ozone or, more logically, column ozone. This quantity is usually stated in Dobson Units (DU) (sometimes, milli-atmosphere centimetres), related to the thickness of an equivalent layer of pure ozone at standard temperature and pressure (approximately at ground level pressure and 15°C); there are 1000 DU in a layer one centimetre thick at STP. Observed values range from 150 to 650 DU. In the tropics, values of column ozone are low, around 220 DU, and show only small, but well defined variations with either latitude or season. Outside the tropics there are such large variations of column ozone from day to day and from place to place that it is necessary to have observations extending over several years before satisfactory average variations can be found. In middle latitudes in both hemispheres the annual variation comprises a maximum in spring and a minimum in autumn. The highest values of column ozone are observed in Arctic regions, where the maximum occurs rather earlier than at lower latitudes. In Antarctic regions the maximum is much later, sometimes occurring only a few weeks before the summer solstice. Peak values are smaller than in the Arctic, rarely exceeding 500 DU. Very recently a new feature has been observed over Antarctica, the development of a deep minimum in early spring (Setpember-October). Values as low as 150 DU have been reported.

These seasonal depletions, the first unequivocal changes in the climatology of ozone to be identified, were completely unexpected, and very divergent opinions as to their cause and their global implications, have been expressed. For this reason, this chapter is devoted to a review of issues well established in the literature, and a separate chapter (Chapter 10) has been given over to the observations of, and hypotheses to explain, Antarctic ozone.

The vertical distribution of ozone in the region accessible to routine balloon ascents is conventionally presented on an ozonagram: an example is shown in Figure 6.1. Position in the vertical is specified by the air pressure, p, using a logarithmic scale; ozone by its partial pressure, p_3, on a linear scale. Also shown are lines of constant mixing ratio of ozone, which serve to define the changes in ozone concentration, at any level, associated with vertical motions. It will be seen that, if the mixing ratio is conserved, air

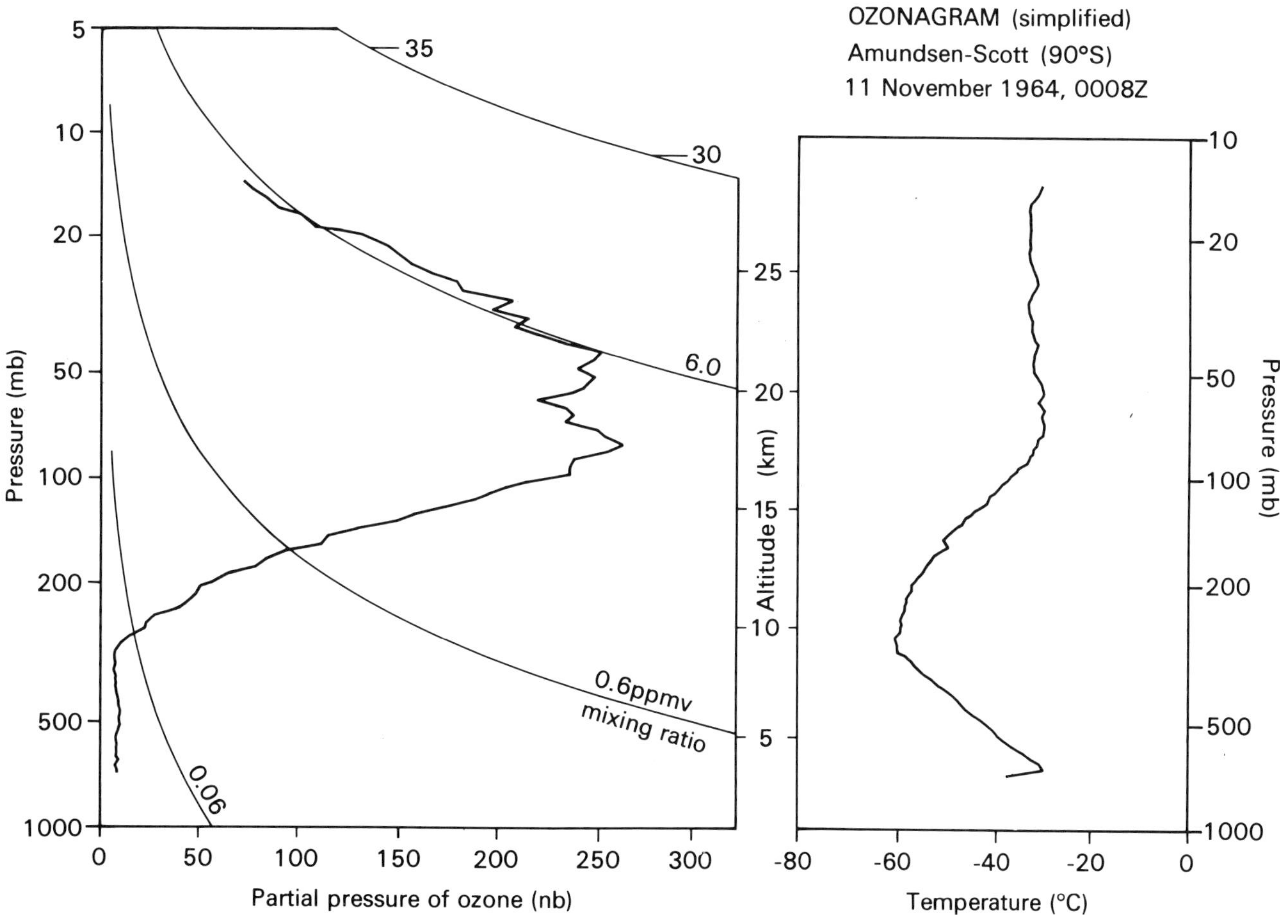

Figure 6.1 Data from an ozonesonde, plotted on a simplified ozonagram, showing altitude profiles of ozone partial pressure (with lines of constant ozone mixing ratio also indicated) and of temperature. The sounding was taken at a late stage in a "final-warming"; the lower stratosphere is very warm. Column ozone was 453 DU

pressure and partial pressure of ozone decrease or increase together. Air pressure is a single-valued function of height, since at any level it depends solely on the weight of air above that level. The equation of state

$$p = nkT$$

relates p to the number density of air, n (molecules cm^{-3}), the temperature, T (Kelvin), and Boltzmann's constant, k. There is a similar equation for each constituent of air, in particular

$$p_3 = n_3 kT$$

defining p_3 in terms of n_3, the number density of ozone. The mixing ratio by volume of ozone is n_3/n, usually stated in parts per million by volume (ppmv), readily pre-plotted since it is clearly equal to p_3/p. An important property of the ozonagram is that the area bounded by the vertical axis, any two horizontal lines and the ozone profile is proportional to the integrated amount of ozone between the pressure levels defined by the horizontal lines. Change in column ozone produced by specified changes in the ozone profile can be readily assessed.

The partial pressure of ozone in the troposphere is small, and often, as in Figure 6.1, changes little with height. The almost abrupt increase of partial pressure at the tropopause (in this figure, the level of minimum temperature) is typical of extra-tropical profiles; in the tropics the rise of partial pressure may not occur until a few kilometres above the tropopause. The most rapid increase in mixing ratio occurs above the level of maximum partial pressure. In this profile the maximum mixing ratio is seen to occur at 20 mb (about 26 km), but it is not well determined. Errors of some 10% are possible in both the pressure and partial pressure measurements at this level. However it is clear that the bulk of the ozone lies well below the level of maximum mixing ratio. Between 100 mb and 30 mb some weak structure can be seen. Such features are usually reproduced in data taken as the sonde falls back to earth after the balloon has burst.

MEASUREMENTS OF OZONE

There are several techniques by which ozone can be measured. From the ground, column ozone and, albeit rather poorly resolved, the ozone profile, can be measured passively using a spectrophotometer. Using laser radar (lidar) good vertical resolution can be achieved. Ozone profiles can be measured from devices carried aloft by balloon or rocket. Most recently instruments on satellites have extended measurements of both column ozone and profiles, to the whole globe.

GROUND-BASED MEASUREMENTS OF COLUMN OZONE

The Dobson ozone spectrophotometer, first developed in 1930, remains the basic instrument in the global network of stations measuring column ozone. It measures the intensity of sunlight at two different wavelengths in the region between 305-340 nm, where there are ozone absorption (Huggins) bands. One of the wavelengths is much less strongly absorbed than the other, and from a knowledge of the ozone absorption coefficient at the two wavelengths, the ozone column density can be deduced. In practice two pairs of wavelengths are used, to reduce the effect of air molecules and aerosol particles. Measurements on direct sunlight yield column ozone amounts with a random error of about 5 DU. Empirical relationships have been derived between readings taken on direct sunlight and on zenith skylight, allowing estimation of column ozone under most weather conditions, with a random error of about 15 DU.

The distribution of Dobson stations is quite uneven: 42 in north temperate latitudes, 11 in the tropics, 7 in south temperature latitudes and 3 and 4 in the Arctic and Antarctic respectively. There are clusters within these groups, with 20 out of 42 northern stations in Europe and 6 out of 7 southern stations in Australia and New Zealand.

Published values of column ozone from stations in the network have been based, since 1957, on a conventional value for a double difference of the absorption coefficient of ozone. Recently derived coefficients may require reassessment of that convention, but no firm recommendation is yet available. Absolute values of column ozone are required for comparison of ground-based and satellite measurements. However, consistent relative values suffice to examine trends in the ground-based data. Considerable effort has been devoted to obtaining improved consistency since 1970, but for only a few stations have attempts been made to eliminate instrumental factors retrospectively from their records. The pre-1970 data have been omitted from most analyses for trends undertaken recently.

GROUND-BASED MEASUREMENTS OF THE VERTICAL DISTRIBUTION OF OZONE

Estimates of the vertical distribution of ozone from measurements of zenith skylight can be made with the Dobson spectrophotometer, using the Umkehr effect. In this technique, the zenith clear blue sky is observed whilst the sun traverses a range of solar zenith angles until just after sunset. Scattering from different altitudes in the atmosphere allows the height profile of ozone to be deduced, in the form of "Umkehr layers". There are nine layers, from layer 1 (6-10 km) to layer 9 (43-48 km). One wavelength pair only need be used, but a wide range of solar zenith angle, 60°-91°, has to be covered.

A new method of reduction, known as the 'short Umkehr', has been tested and accepted. It requires about one-third of the observing time needed for the conventional Umkehr. This economy is achieved by reducing the range of zenith angle to be covered to 80°-89°, replacing the information previously gained at lower zenith angles by readings taken on two additional wavelength pairs.

Most Dobson spectrophotometer stations have made occasional Umkehr measurements, but since the technique depends on having clear skies for several hours, the number of observations completed varies greatly from month to month and from station to station. Only some 12 stations have series with enough data to permit trend analyses, 3 in Europe, 3 in North America, 3 in Japan, 2 in India and 1 in Australia.

The differential absorption laser (DIAL) technique has been widely applied to measurements of trace gases in the boundary layer, using long paths. Recent technological advance has made available powerful tunable lasers which allow the vertical sounding of some minor constituents, most notably ozone. A ground-based UV lidar system has been developed in France using the DIAL technique for monitoring ozone number density profiles under clear sky conditions. Dual wavelength operation is required, with a wavelength separation not exceeding 5 nm, to permit retrieval when aerosol layers are present in the troposphere or lower stratosphere. It is further necessary to use different wavelength pairs for retrieval below and above the ozone peak, to achieve acceptable error limits and rate of acquisition. The full sequence of a sounding is computer controlled, and a real-time analysis of the data can be performed. This is the first operational system capable of measuring the vertical distribution of ozone from the ground to 40 km. Other systems, with more restricted capabilities, are being developed in Japan, Germany and the USA.

BALLOON-SONDES AND ROCKET SONDES

By 1979 some 8000 balloon soundings of ozone had been made, and attempts to prepare a world-wide ozone climatology undertaken. The set was far from homogeneous, the numbers and dates of the soundings and the technique used have varied drastically from station to station. The geographical distribution was extremely irregular with a strong concentration on northern mid-latitudes. The coverage in the tropics and over most of the Southern Hemisphere was very poor. There were however, data from Antarctica for 1964-1966.

Following the development of satellite instruments, the number of stations launching ozone-sondes to a regular schedule fell sharply. Emphasis has shifted towards obtaining improved accuracy, for verification of satellite data, and to specific case studies, such as the Antarctic ozone depletions. Several new sensors have been developed and tested.

The conclusion from three campaigns conducted in 1983/4 to compare balloon-borne ozone sensors, was that most sensors agreed with one another within about ±10% between about 20 km and 40 km. This was the case both for the small EEC ozonesondes and for the larger in-situ UV photometers. Further laboratory tests confirmed this agreement, except that in principle the UV photometers could be accurate to better than ±5%. As part of the same campaign, measurements by balloon-borne sensors were compared to Umkehr soundings and those from satellite instruments (SBUV on Nimbus 7; see below). The level of agreement was also within about ±10% at all altitudes.

Generally speaking, rocket sondes have been used in more recent years for case studies of mesospheric and upper stratospheric ozone rather than launched on a regular schedule as part of a network.

MEASUREMENTS FROM SATELLITES

A notable recent achievement has been the analysis, validation and release of data obtained by instruments flown on the Nimbus 7, Applications Explorer II and Solar Mesospheric Explorer satellites. Ozone profiles have been retrieved by three independent systems, using measurements in the ultra-violet (SBUV/TOMS), visible (SAGE) and infra-red (LIMS) spectral regions, respectively. The ultra-violet measurements yielded also column ozone. A summary of the techniques used and a list of the other constituents and parameters retrieved is given in Volume II of NASA/WMO, 1986. (Special Introduction to Chapters 8, 9, 10 and 11.)

Improved versions of two of these measurement systems are already operational, SBUV-2 on the NOAA satellite series and SAGE-2 on the Earth Radiation Budget Experiment (ERBE) satellite, SBUV-2 performs daytime retrievals of column ozone and of vertical profiles of ozone above the ozone peak (25-55 km approximately). Instrumental drift is checked by an on-board calibration lamp. If this system performs to its specification it will be able to detect changes within the profile of about 1.5% per decade. SAGE-2 has 7 channels in the visible, and profiles of O_3, NO_2 and H_2O, and of aerosol extinction, are retrieved.

MEASUREMENTS FROM THE SPACE SHUTTLE

Stability is a major concern with satellite instruments required to operate continuously for several years. The problem is acute when the aim is to detect a slow trend in the quantity observed. Regular flights of a similar instrument for a few orbits in the space shuttle, with careful laboratory checks before and after each flight, could provide better control than has been available previously. A shuttle-borne SBUV (SSBUV) instrument has been built to serve as a standard for the current operational SBUV-2 system. The plans to deploy it were disrupted by the disaster of January 1986, but flights of the SSBUV have high priority when the shuttle programme can be resumed.

FLUCTUATIONS AND TRENDS IN GLOBAL COLUMN OZONE

Column ozone is highly variable over a wide range of time scales. At individual stations in mid-latitudes, day to day changes are often large; the monthly range of daily mean values frequently exceeds the annual range of monthly means. The more conspicuous of these changes are clearly related to shift or changes in intensity of stratosphere flow patterns (high ozone values in upper air troughs, low values in the ridges). Local monthly mean values may thus vary considerably, successively and from year to year, in response to fluctuations in the general circulation.

The effects of some of these fluctuations can be eliminated by taking the global average of column ozone. The top panel in Figure 6.2 shows monthly mean values of this average obtained from the network of Dobson spectrophotometers from 1958 to 1985. Prior to 1958, the network was too sparse to be representative. The bottom panel shows the global average of column ozone calculated from data obtained between November 1978 and September 1984 by the SBUV instrument aboard the Nimbus 7 satellite. Both time series have been filtered to remove systematic seasonal variations. The vertical scales indicate percentage deviations from the overall averages after removing the seasonal patterns.

The obvious question is: has there been any consistent trend in global column ozone? A visual inspection would suggest that, certainly up to 1982, any trend was masked by the large variability. Indeed although several statistical analyses have been carried out on the global column ozone time series up to that year, no consistent figure for a trend can be agreed.

It can be seen that, after 1982, two very large excursions have occurred, and some analyses which have been performed up to 1983 do show a significant downward trend. However these analyses have undoubtedly been heavily influenced by the deep minimum of 1982/83, which has not persisted.

The estimated change in the global average column ozone derived from ground-based (Dobson) data over the period 1970-1985, is a decrease of 0.05 ± 0.09% per year (Chemical and Engineering News, 1986). This is not statistically significant. The satellite data between 1978 and 1984 yield a very different result. There is no apparent recovery from the low values of 1982-83, and an average decrease of about 0.6% per year over the six years is indicated. However, the abrupt cessation of accord between the satellite and the ground-based data in mid-1983 has not

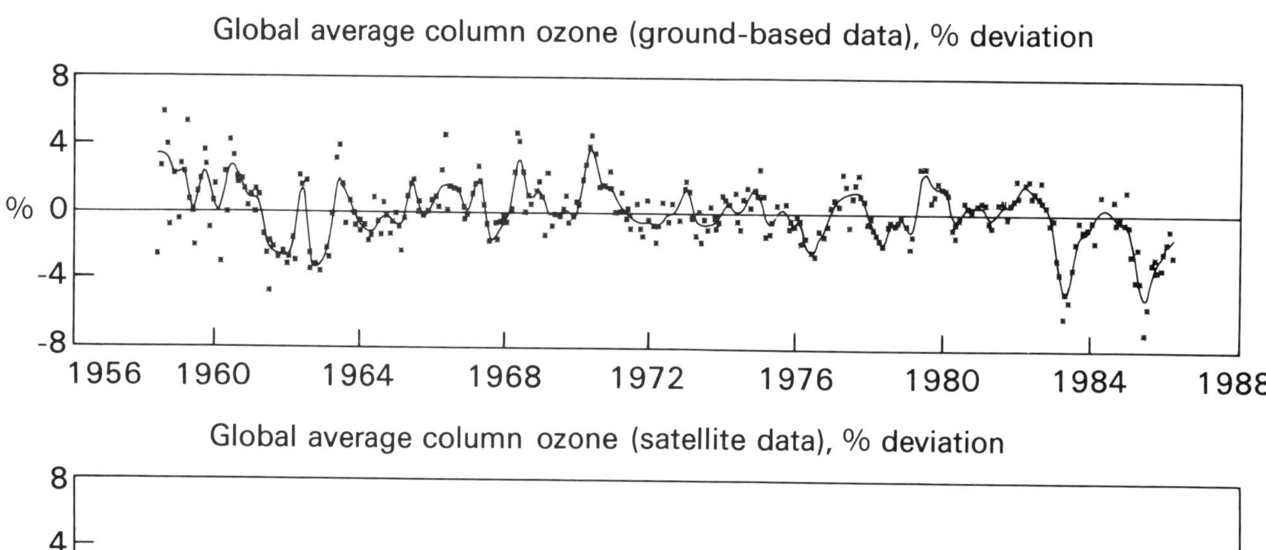

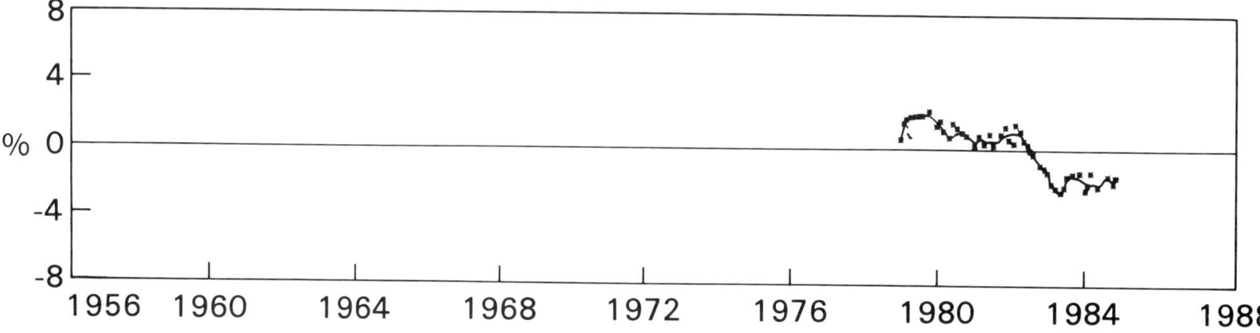

Figure 6.2 Monthly mean values of globally-averaged column ozone obtained from the Dobson network (1958-1985, top panel), and from the SBUV instrument on Nimbus 7 (November 1978-September 1984, bottom panel). Both time series have been deseasonalised; percentage deviations are from the overall averages (after Chemical and Engineering News, 1986)

yet been adequately explained, and no firm conclusion can yet be drawn. It should be noted that the large decreases in Antarctic ozone which are discussed in detail in Chapter 10, do not significantly affect the long period time series of global mean column ozone, because they are limited to a relatively small area and to only one or two months of the year.

The only conclusion which can be made about long term trends in globally-averaged column ozone, is that the large temporal variability precludes the identification of any systematic trend. Although there may be an underlying trend since about 1980, it is difficult to discern with confidence because of the particularly large variability in the last five years.

THE 1982/83 MINIMUM IN COLUMN OZONE

Attempts have been made to link the 1982/83 minimum of column ozone to volcanic eruptions and/or changes in sea-surface temperatures which occurred earlier in 1982:

A. The eruptions of the volcano El Chichon (Mexico) in March-April 1982 injected large amounts of particulate matter, and possibly of gaseous material as well, into the tropical lower stratosphere. This could have perturbed the chemistry in that region, either by reason of the greatly increased particulate area available for heterogeneous chemistry, or by the direct injection of, for example, hydrochloric acid (HCl). However, the radiative balance of the tropical lower stratosphere was strongly perturbed; increases of some 5 K in monthly mean temperatures, persisting for several months, have been reported. Consequent changes in the diabatically forced circulation linking lower and middle latitudes could account for much of the observed variation in total ozone.

B. The 1982/83 period was, coincidentally, that of a major EL Nino even (a rapid change in the surface temperature of the equatorial Pacific Ocean), beginning shortly after the eruptions, and to which some large tropospheric climatic anomalies have been attributed.

The difficulty of separating the effects of El Nino from those of El Chichon on the temperature in the extra-tropical stratosphere has been specifically demonstrated. There will be similar difficulty in analysing the variations in ozone amounts.

TRENDS IN THE OZONE PROFILE

The most important result on trends within ozone profiles comes from a study of data from the Umkehr stations. The analysis indicates statistically significant trends in the upper Umkehr layers, 7 and 8 (33 to 43 km altitude), of −0.2 to

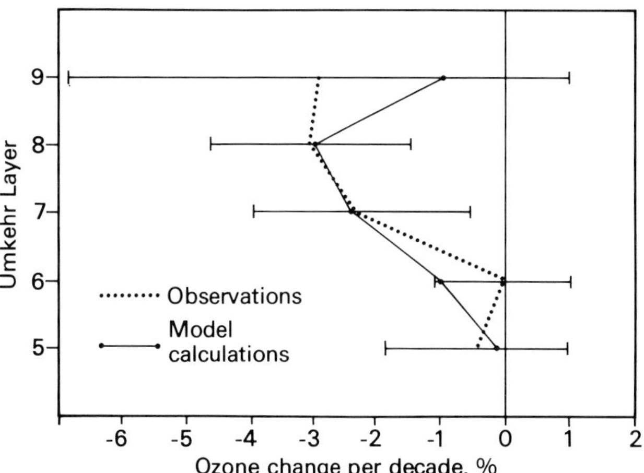

Figure 6.3 The rate of change of ozone concentration (percent per decade) between 1970 and 1980 at five levels from 23 to 48 km. There is substantive agreement between these observations and the model calculations of Wuebbles. (Lawrence Livermore National Laboratory)

−0.3% per year over the period 1970-1980, with little trend in the lower layers, 5 and 6. There is substantive agreement between these results and calculations using one-dimensional models (the continuous line), as shown in Figure 6.3. Two points have, however, to be considered, which in combination will enjoin caution in accepting this agreement at definitive. Firstly, the results (including the sign) are very sensitive to the corrections applied for the effects of aerosols. In the analysis quoted, the first to include such corrections, the aerosol concentrations measured at Mauna Loa (Hawaii) were assumed to exist globally, an approach with obvious limitations. Moreover, attempts to extend the analysis into the period affected by the eruptions of El Chichon (see above) suggested that considerably more work would be required to enable the Umkehr data for that period to be used with any confidence.

The second point is that these results come from just 13 stations, unevenly distributed, only one of which is in the Southern Hemisphere. This sampling problem has been examined by comparing the Umkehr trends with trends derived from 42 months of SBUV data. It appears that the station average is quite close to the SBUV area-weighted trend, but this agreement could well be fortuitous. Trends at individual stations are not, in general, representative of trends in appropriate zonal averages of the SBUV data. Thus, the overall result could be very sensitive to the availability of stations.

NEW PERSPECTIVES FROM SATELLITE DATA

Although the TOMS and SBUV data sets are not yet of sufficient length to examine decadal trends, they have already provided valuable new perspectives on the distribution and variation of ozone, and have made it possible to

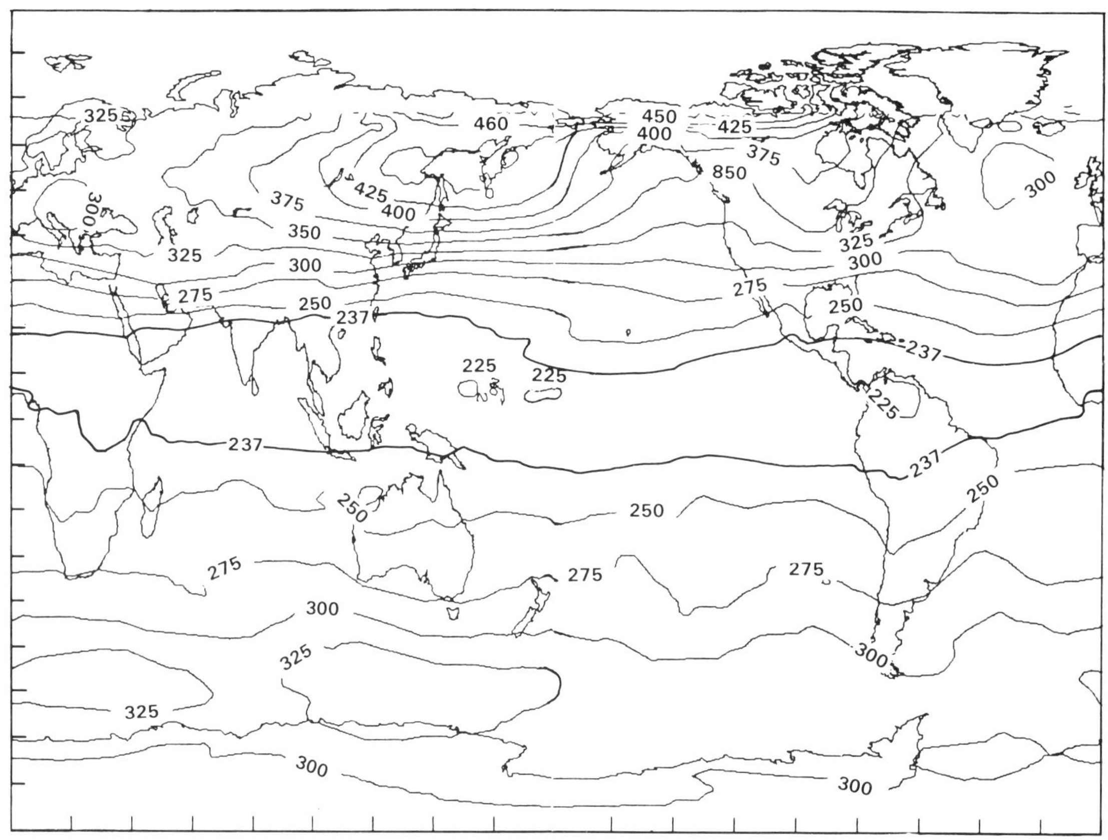

Figure 6.4 Monthly mean values of column ozone (DU) for January 1979, from Nimbus 7 TOMS instrument

assess the extent to which earlier findings have been biassed by the irregular distribution of the ground-based network. Figure 6.4 shows the average column ozone for January 1979 as derived from TOMS data. Considerable variation of ozone with longitude can be seen. In the equatorial belt of low column ozone, there are minima over the western Pacific and over Central America. In the winter (Northern) hemisphere there are maxima at high latitudes, over eastern Siberia and Arctic Canada, whilst in the summer (Southern) hemisphere there are maxima at about 55°S, south of Africa and south of Australia.

Figure 6.5 shows a four-year average of ozone mixing ratio in January as a function of height and latitude, derived from SBUV data. As expected for a photochemically produced species, the mixing ratio is highest at sub-solar latitudes in mid-stratosphere. However (as was shown in Figure 6.1) the greatest number densities are found much deeper in the atmosphere. The distribution of the mixing ratio below 30 km is highly informative. The belt of high column ozone in the Southern Hemisphere (Figure 6.5) is associated with mixing ration gradients directed to higher latitudes in the tropical lower stratosphere, a clear indication that transport is playing an important role. A comparison between Figure 6.5 and computed profiles is given in Chapter 8. There are disconcertingly large discrepancies in the height range where photochemical processes had been expected to be dominant.

Figure 6.6 shows ozone on the 850 K potential temperature surface (30 km, 10 mb) for one day in January 1979. Air of low ozone content from the polar vortex is being extruded round the southern flank of the Aleutian anticyclone, whilst a tongue of ozone rich air from low latitudes is being drawn over the pole from about 120°E, the region of confluence of the cyclonic and anticyclonic

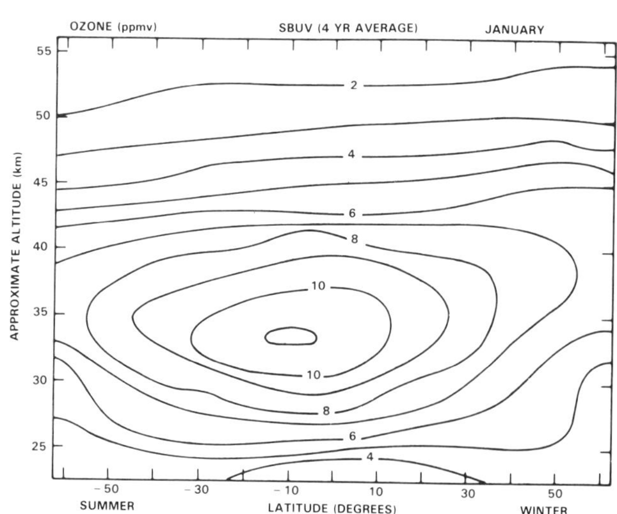

Figure 6.5 Four year average January ozone mixing ratio as a function of latitude and altitude, from the SBUV instrument

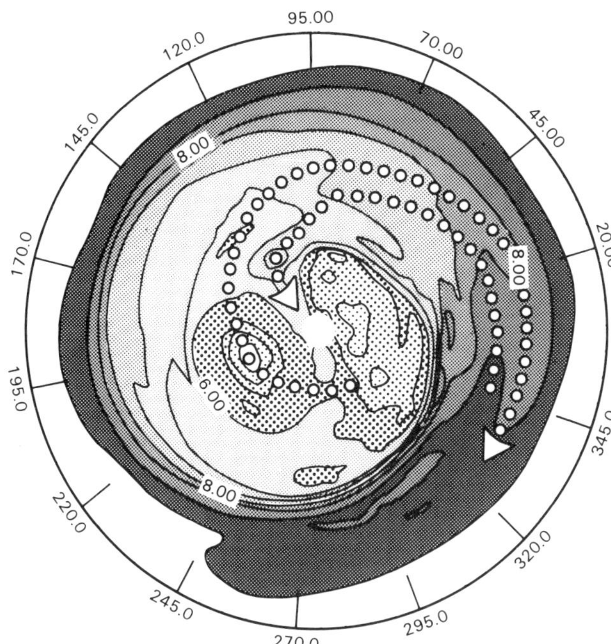

Figure 6.6 Ozone mixing ratio (ppmv) on the 850K isentropic surface (10 mb, 30 km), 27 January 1979, from LIMS data. The arrowed lines indicate the approximate paths of ozone-poor air from the polar vortex, and ozone-rich air from low latitudes

flows. This vividly illustrates the value of the global perspective provided by satellite data for diagnosing the actual transport processes.

OZONE IN THE TROPOSPHERE

In early attempts to estimate the global budget of ozone it was assumed that ozone in the troposphere was completely passive; it was generated in the stratosphere, injected through the tropopause and mixed downward to be destroyed by contact with various substances, mainly organic. Thus in this theory the surface was a sink, the strength of which was identical with the influx through the tropopause. In the 1950s it was recognised that the high concentrations of ozone found in the polluted air of certain big cities in specific meteorological conditions (smog episodes) were due to local production of ozone resulting from photochemical reactions involving nitrogen oxides and volatile organic compounds. However, it was thought that such infrequent and highly localised events had no global significance. Subsequent studies of the processes related to the oxidation of methane made it clear that ozone was not inert even in a clean troposphere. The processes are complex and can lead to net production or net loss of ozone according to the amount of NO_x present. A more detailed explanation is provided in a companion report (PORG, 1987).

The mechanisms by which ozone is transported into the tropsophere, and subsequently into the boundary layer, are poorly understood. It has been shown that the influx of stratospheric air into the troposphere is not a smooth and continuous process, but rather occurs in discrete incursions; some related atmospheric structures are 'cut-off lows' and 'tropopause folds', the latter mainly associated with jet stream development (see Chapter 5). The transport of ozone from the lower troposphere into the boundary layer is strongly dependent on local topography and meteorological conditions, and is subject to large daily variation.

Studies based mainly on ozonesonde data show that surface ozone at mid-latitudes displays two modes of seasonal behaviour: a broad maximum in summer is seen in densely populated and industrialised regions, such as Europe and the United States, whereas there is a maximum in spring in sparsely populated regions remote from industrial activity, such as Canada and Tasmania. This is consistent with an earlier findings that the tropospheric ozone content is lower in the Southern than in the Northern Hemisphere, a finding which remains tentative due to the scarcity of observations in the Southern Hemisphere. The seasonal cycle of ozone in the middle troposphere over Europe, the United States and northern Japan is similar to that at the surface, with a maximum in summer. This is quite different from the cycle at 300 mb, in the upper troposphere, which shows a maximum in spring similar to the cycle in column ozone. There are indications from the current data base in combination with limited historical data that the mixing ratio of ozone near the surface in rural areas of Europe and in the central and eastern United States may have increased by about 10-20 ppbv (50-100%) since the 1940s. This seems to have been accompanied by an increase of ozone in the middle troposphere over Europe in the last 15 years, and perhaps also over northern America and Japan. It has been suggested that the summer maximum and the observed trends arise mainly from photochemical production associated with emissions of oxides of nitrogen, hydrocarbons and carbon monoxide from combustion of fossil fuels. However, with results from only a limited number of ozonesonde stations it remains questionable as to whether the increase of tropospheric ozone reflects the increased production of ozone there, due to increased man-made emissions, or due to increased transfer from the stratosphere.

SUMMARY

1. The large temporal variability in globally averaged column ozone precludes the identification of any systematic trend.

2. Analysis of ozone profiles indicates statistically significant trends between 30 and 40 km, of between –0.2 to –0.3% per year between 1970 and 1980, with little trend at lower altitudes.

3. Perhaps the most significant observational development has been the recognition of the spring depletions of ozone in Antarctica. The observed changes are so large and so rapid that significant advances in our understanding of stratospheric behaviour may be expected when intensive measurements of trace gases are undertaken there.

MEASUREMENTS OF OTHER STRATOSPHERIC CONSTITUENTS 7

It is the purpose of this chapter to describe shortcomings and recent progress in measurements, to recommend what measurements should be given immediate priority, and to show how close we can expect to come to ideal measurements of trace gases in the future. These constituents have very low concentrations and are often very difficult to measure, so that the quality of present measurements is frequently less than ideal. The constituents important to ozone chemistry are listed in Table 7.1, which also highlights the status of the measurements. Figure 7.1 shows approximate profiles of some of these constituents. Note the contrast between the shapes of the profiles of most source gases which are transported from the troposphere (CH_4, N_2O, CFCs) and those of radicals (NO, NO_2, ClO) which are created in the stratosphere.

The availability of high quality measurements of stratospheric composition is important for our ability to test models of the stratosphere. These models should correctly represent the chemistry and physics of the gases which control the ozone balance, their distribution around the globe, and their changes with time of day and with season; measurements should test the accuracy of this representation.

Ideally, then, most constituents should be measured with high accuracy at a variety of heights, latitudes, seasons and times of day. This would argue for measurements over the whole globe, and therefore exclusively for measurements from satellites. However this is not necessarily the case, and this point is discussed below in more detail; furthermore, not all measurement techniques can be used from satellites. Figure 7.2 depicts the variety of platforms and techniques used so far.

Measurements of several important constituents continue to be inadequate, or non existent:

(a) The predicted existence of HOCl, and HNO_4 has not yet been positively confirmed by measurements (although strong evidence for the eixstence of HNO_4 is reported). The predicted profiles and diurnal variations of other reservoirs (N_2O_5, $ClONO_2$ and H_2O_2) have not yet been confirmed. Although measurements of reservoir gases have been greatly improved recently by results from the ATMOS instrument, they are still far from able to constrain models.

(b) The measured changes of NO_2 at night are consistent with model predictions, but are not sufficiently accurate to confirm the details of N_2O_5 chemistry in models.

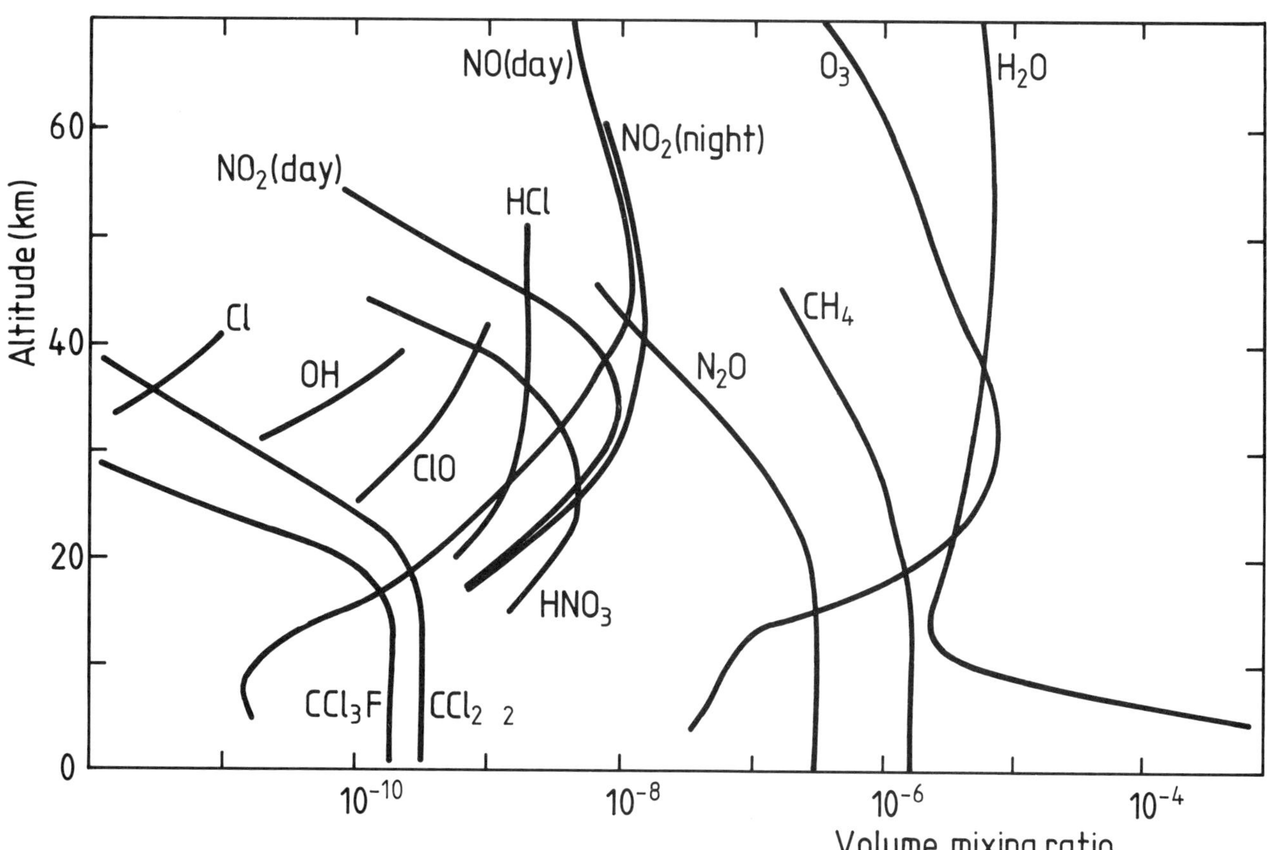

Figure 7.1 Probable mixing ratios of some of the atmospheric constituents relevant to ozone. Some of the measurements of these constituents have large uncertainties — this figure is meant only as a rough guide

(c) The ratio of O to O_3 is a basic prediction by models, but there have been only a few measurements of O (atomic oxygen). On only one balloon flight was ozone measured simultaneously, and doubt has been cast on the accuracy of that measurement of ozone.

(d) OH has been measured infrequently by few independent techniques. There are no OH measurements below 30 km and at 35 km the different techniques differ by a factor of 20. Only one technique has measured HO_2 below 25 km and the results differ from model predictions by a factor of 10 in the lower stratosphere. The ratio of OH to HO_2 is an even simpler calculation in models yet differs from the ratio of these measurements by a factor of at least 30.

(e) Although measurements of ClO by remote sensors do not differ greatly from those by local samplers, the variability of the sampling measurements is very much greater than that of the remote ones.

On the other hand, during the last 10 years, models have been constrained or predictions have been confirmed by many important discoveries from satellite data:

(a) The fall in concentration of CH_4 between 30 and 40 km is matched by a rise in H_2O over a wide range of latitudes from January to May 1980. The rise in H_2O is consistent with predictions of CH_4 oxidation within 10%.

(b) Near equinox, the North-South distributions of CH_4 and N_2O have a double peak, quite unexpectedly. As discussed in Chapter 8, some 2-D models can now predict this behaviour and Figure 7.3 shows some examples.

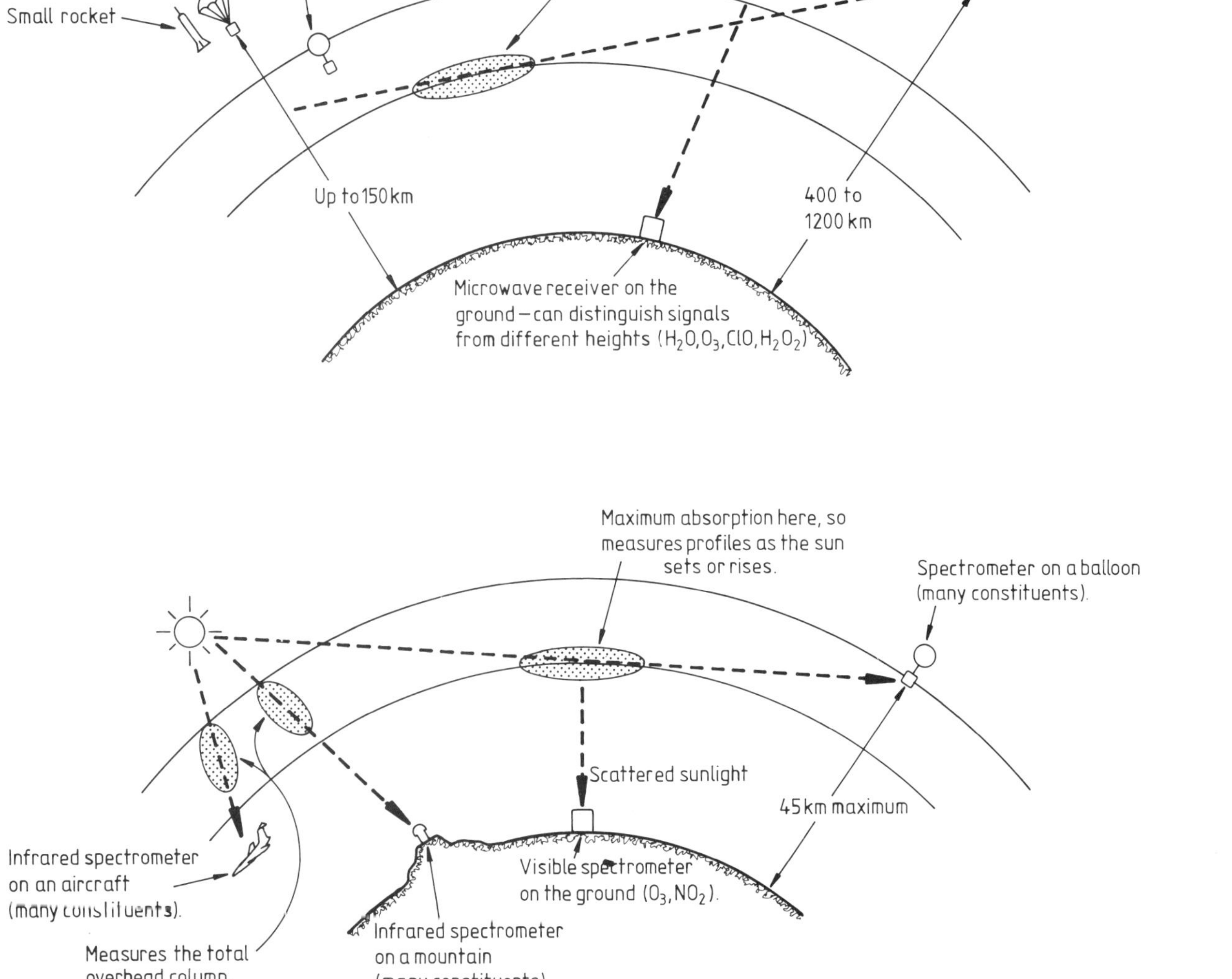

Figure 7.2 Some of the many techniques and platforms for measuring minor constituents in the stratosphere. Balloons also often carry radiometers which sense thermal emissions, and satellites have carried spectrometers which measure solar absorption

(c) The measured vertical profile of NO_2 near the winter poles has confirmed that there is a source of reactive nitrogen in the winter thermosphere which is transported downwards. Such a source was first predicted only in 1983, and may conttribute to the global budget of reactive nitrogen in the stratosphere.

(d) The sudden step in NO_2 concentrations near 55°N in winter over Canada under certain conditions (the "Noxon cliff") is linked to dynamical events. Both the LIMS and SAGE satellite sensors showed that the polar air at some longitudes has a low concentration of NO_2 when the vortex is displaced.

(e) Close to sunset, the change of NO_2 with solar angle was measured by LIMS during May 1979 from 56°N to 74°N. The results confirm model predictions rather well, except that there is potential ambiguity in the trend at night because the measurements at different local times are also on different days spread over a month, and NO_2 was changing during the month. In practice, this change occurred only at the higher latitudes which did not affect the trend at night.

(f) Measurements from LIMS, when analysed by following parcels of air along trajectories, show that there is a significant source of HNO_3 at high latitudes in winter. Models do not predict this source, and this is described in more detail in Chapter 8.

(g) Spectra from ATMOS, flown on the Space Shuttle, have confirmed the existence of chlorine intrate ($ClONO_2$), predicted since 1976 but not observed until now; and have provided excellent new evidence for the existence of N_2O_5 and HNO_4.

(h) The 'hygropause', or minimum in mixing ratio of H_2O about 2 km above the tropopause, measured locally from balloons, had been shown from LIMS measurements to be a persistent feature at low latitudes. An hypothesis associated with equatorial thunderclouds has been proposed to explain this phenomenon, but the necessary aircraft measurements to confirm it will not take place until 1987.

(i) Aerosols have been detected in the stratosphere, and in unexpectedly large concentrations at high latitudes (polar stratospheric clouds). These may have an important effect on stratospheric chemistry.

However there is a major problem with many of these remote measurements — that the absolute calibration of

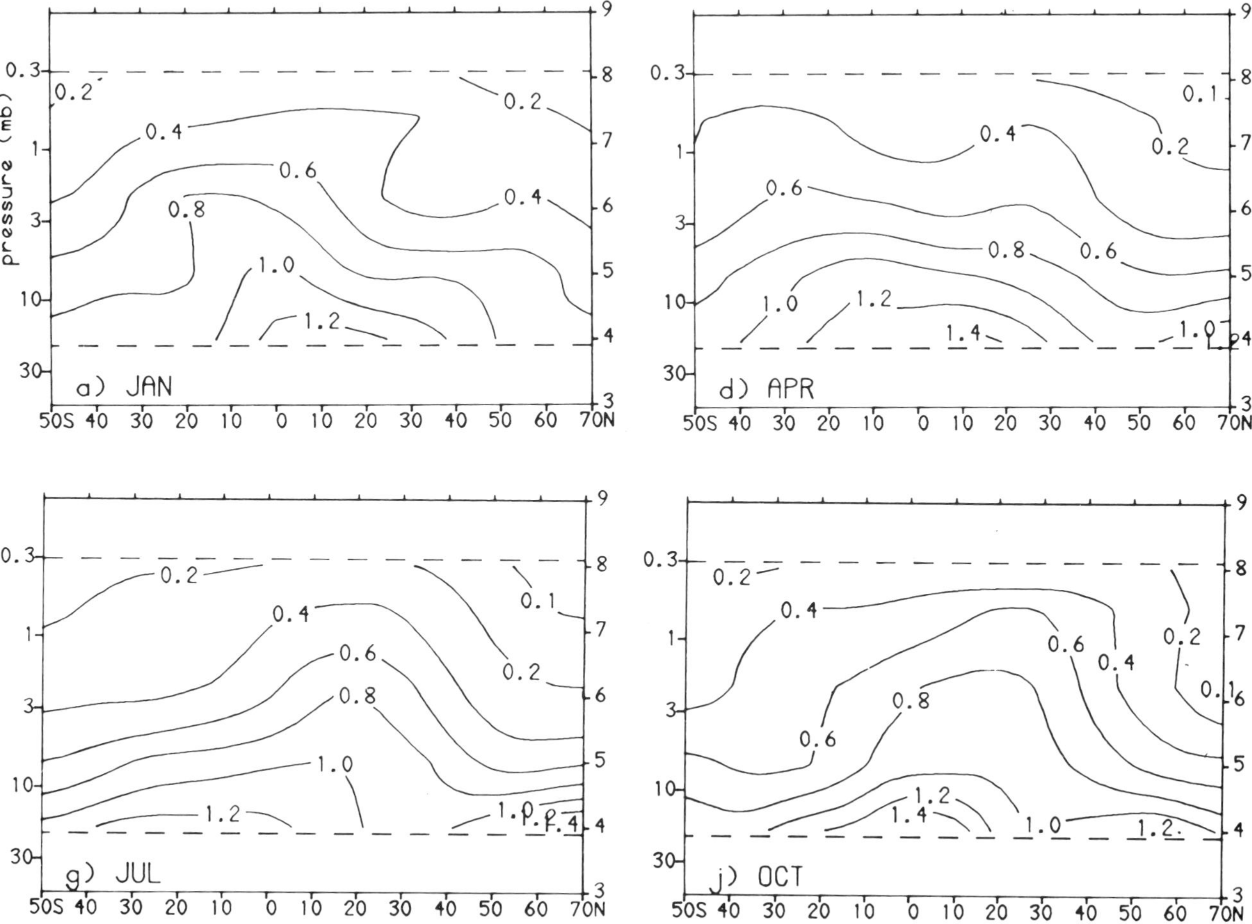

Figure 7.3 Distributions of CH_4 and N_2O, averaged around latitude circles, measured by the Stratospheric and Mesospheric Sounder (SAMS) on the Nimbus 7 satellite.

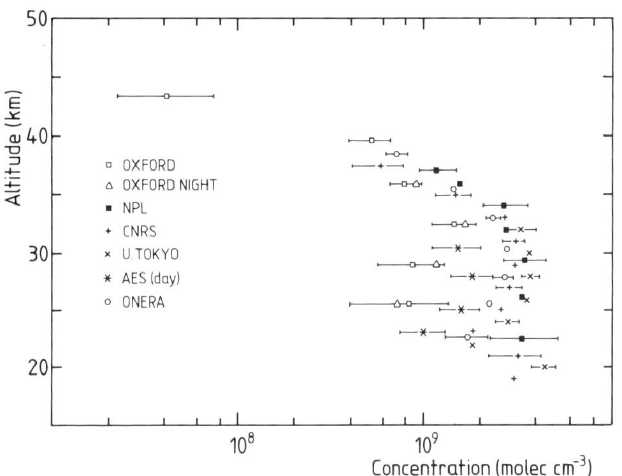

Figure 7.4 NO_2 profiles measured during the second Balloon Intercomparison Campaign (BIC) in June 1983, from Palestine, Texas. Note the wide spread in results despite the near-simultaneity of the measurements

some of the sensors is in significant doubt. This has been shown forcefully by the 1982 and 1983 Balloon Intercomparison Campaigns, where many remote sensors, including ones similar in principle to the satellite sensors, measured the same constituents simultaneously. Measurements of NO_2 below 30 km differed by a factor of 3 (Figure 7.4), and these differences cannot be reconciled (almost certainly they are associated with the line shape of NO_2 in the infra-red, and with the temperature dependence of its visible spectrum). Above 30 km, measurements of HNO_3 differed by ±40%. Although of the remote measurements H_2O has probably been measured to the best accuracy, even this is only ±25% (much of this is probably due to errors in the infra-red line shape of H_2O). These uncertainties restrict the usefulness at the present time of the satellite and balloon data. (For example, the uncertainties in H_2O and CH_4 calibrations means that it is just possible that the consistency in (a) above could be fortuitous.) Future improvements to spectroscopic data should reduce these uncertainties.

There is a similar problem with chemiluminescent measurements of NO. Between 30 and 35 km, balloon-borne samplers measure typically twice the NO measured by outwardly similar samplers on rockets. Recent flights have shown that the difference almost certainly lies in the simpler on-board calibration of the smaller rock-borne sensors, but until this point is resolved, typical NO concentrations must remain uncertain. Coupled with the uncertainty in NO_2 sensors some workers would argue that the total reactive nitrogen in the mid-stratosphere must be considered uncertain within at least a factor of 2.

Technical problems with SAMS on Nimbus 7 satellite meant that there was no NO data, and that the calibrations of the CH_4 and N_2O data in the lower stratosphere were in error. The signal-to-noise ratios of SAMS meant that most of the data had to be averaged around latitude circles and over one month but, given the status of models, useful comparisons could still be made. Sensors on UARS should avoid these technical problems and should fill many gaps in the global measurements of constituents, in particular of NO, HCl, HF and CFCs. Figure 7.5 depicts the constituents and instruments of UARS, which will be launched in 1991. These instruments should have much better signal-to-noise ratios than previous satellite sensors, and by the early 1990s 3-dimensional models should contain enough chemistry to be able to make useful comparisons with the measurements.

THE NEED FOR SIMULTANEOUS MEASUREMENTS

Frequently models can predict the ratio of a pair of constituents with rather more apparent certainty than their absolute concentrations. For example, the concentration of NO_2 at 25 km depends on those of O_3, N_2O, HNO_3, O, NO and on the photolysis rate coefficient for NO_2; whereas the ratio of NO to NO_2 depends only on O_3 and the photolysis rate coefficient for NO_2 to a first approximation. A simultaneous measurement of the three constituents, together with temperature and the solar flux, would then provide an important test of model predictions. This argument has frequently been applied to joint measurement campaigns from balloons, as well as to UARS. It must remain an important future goal, but unfortunately the present uncertainties in the absolute calibrations of many sensors means that the total uncertainty in the measurement of a ratio frequently exceeds a factor of two — these uncertainties are not usually equal for each constituent. Furthermore, the variability of some constituents is now being shown to be small at restricted latitudes and seasons (eg mid-latitudes, equinox), in many cases less than the error of the measurements. When this is the case, it is obvious that there is no point in simultaneous measurements. When, in the future, the error on these measurements is reduced because of improvements in techniques or calibrations, simultaneous measurements will become important.

OPERATIONAL SATELLITES AND THE NEED FOR NON-SATELLITE MEASUREMENTS

For many constituents, there has been a progression of measurement techniques. Usually, the first detection was by balloon-borne sensors (though sometimes by spectrometers in aircraft or on the ground); then balloons were flown at several sites and seasons with a variety of sensors, aircraft flew to all latitudes and sensors on the ground measured over a range of seasons; finally remote sensors were deployed on satellites to give global coverage so that such interesting features as, for example, the double-peaked structure of CH_4 and N_2O, could be observed. Measurements of some constituents are only at the first or second of these three stages, but progression to the satellite stage is probable in the future. However, satellites are not neces-

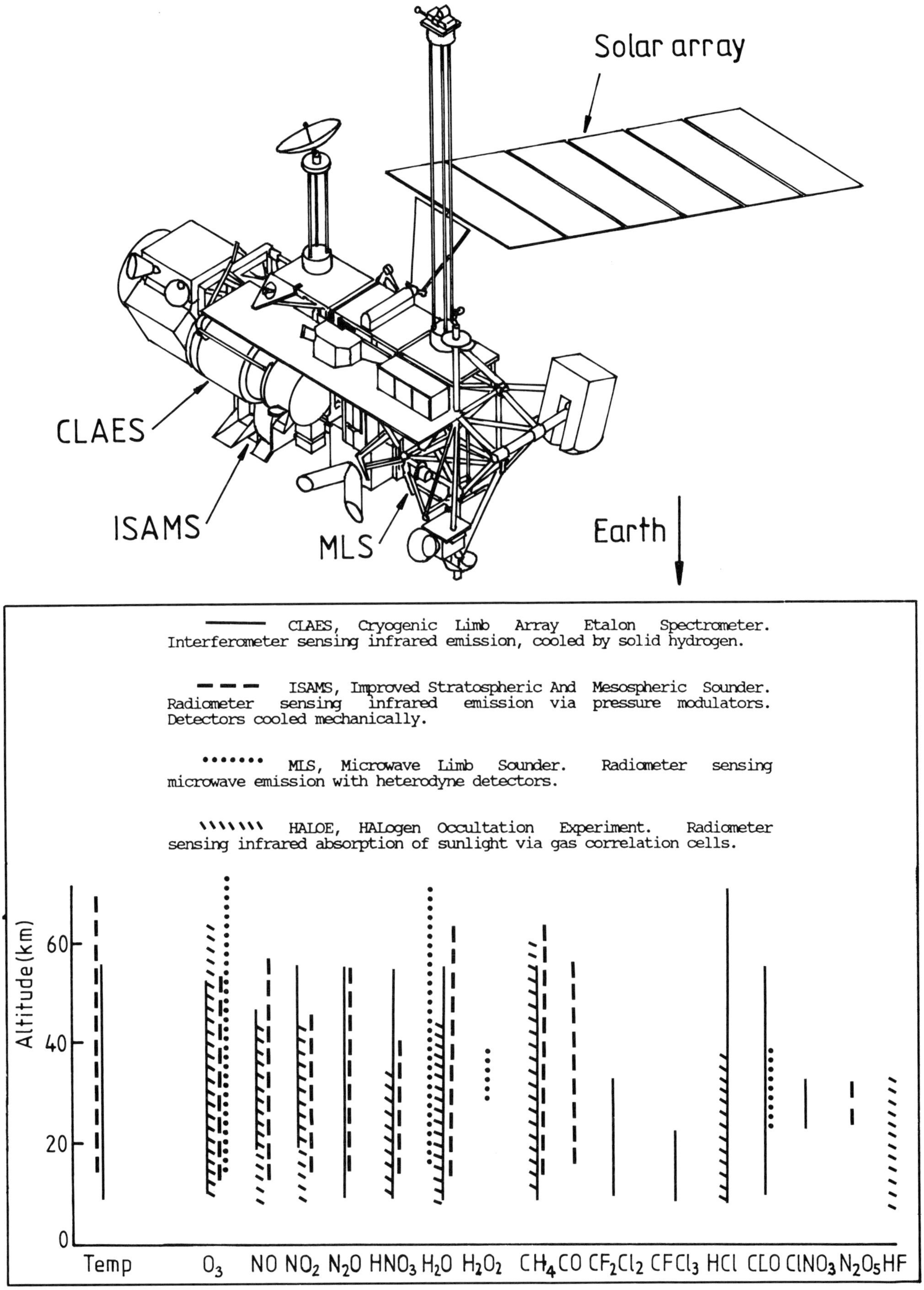

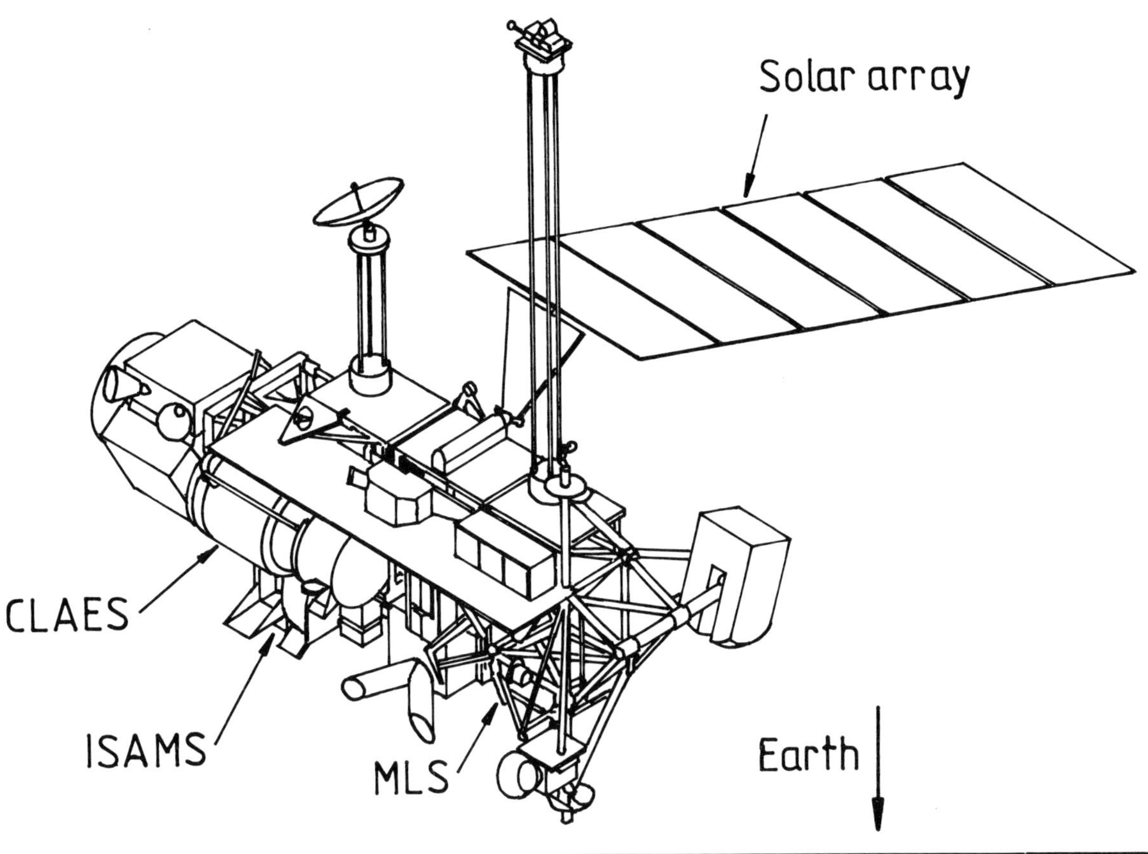

——— CLAES, Cryogenic Limb Array Etalon Spectrometer. Interferometer sensing infrared emission, cooled by solid hydrogen.

— — — ISAMS, Improved Stratospheric And Mesospheric Sounder. Radiometer sensing infrared emission via pressure modulators. Detectors cooled mechanically.

••••••• MLS, Microwave Limb Sounder. Radiometer sensing microwave emission with heterodyne detectors.

\\\\\\\\ HALOE, HALogen Occultation Experiment. Radiometer sensing infrared absorption of sunlight via gas correlation cells.

Altitude (km)

60
40
20
0

Temp O_3 NO NO_2 N_2O HNO_3 H_2O H_2O_2 CH_4 CO CF_2Cl_2 $CFCl_3$ HCl ClO $ClNO_3$ N_2O_5 HF

Figure 7.5 NASA's Upper Atmospheric Research Satellite (UARS), to be launched in 1989, will carry four new sensors which will measure minor constituents in the stratosphere

sarily the optimum platform even for remot sensors, and several disadvantages offset the obvious advantage of satellites' global coverage:

(a) They have a low cost per data point, but other platforms are cheaper if the aim is mere detection (eg HBr) or measuring a single profile (eg $ClONO_2$).

(b) Diurnal variations are much harder to measure than from other platforms, except from a geostationary satellite which introduces severe technical problems because of its distance from the earth.

(c) The vertical resolution is often worse than for balloon-borne sensors since the distance of satellites from the atmosphere requires a much better telescope with a much narrower field of view for equal resolution.

(d) Remote measurements of some radicals are difficult and require long integration times (eg OH). Until these sensors are further developed, they are best deployed on balloons where the same air parcel can be viewed for many hours. This deployment will, of course, hasten improvements in the sensors.

(e) Satellites have strictly limited life and stable calibrations are difficult to maintain. All attempts to measure small changes with time must ultimately rely on measurements from the ground, if only to confirm the stability of measurements from satellites, unless the satellite instrument can be retrieved and recalibrated in the laboratory. (Note that this should be possible in the future, with NASA's concept of recoverable payloads.)

This last difficulty with satellite sensors is largely removed by the concept of operational satellites. Here, several nominally identical sensors are constructed, and are calibrated at the same facility. They are then mounted on successive satellites so that there is always one or more aloft. Their calibrations are compared frequently to those of sensors on the ground or on balloons. So far, only temperature and ozone have been measured operationally, but there are plans for instruments to measure other constituents from future operational satellites.

The most accurate measurements of most source gases (CH_4, N_2O, CFCs) are by laboratory analysis of cryogenic or whole-air samples, collected from aircraft or balloons. Unfortunately these flights cannot give complete global coverage, so that satellite sensors are essential to guarantee that these flights are representative of the globe. Of these source gases, only CH_4 and N_2O have been measured from satellites so far, and those not in the lower stratosphere; again sensors on UARS should remove this major shortcoming.

An important new technique for comparing model predictions with measurements is to follow measurements of constituents along trajectories (air parcels). So far, this has been done with data from LIMS where only a few measurements are made each day at each location and the signal-to-noise ratio at each point is limited. A better concept, not yet developed, would be to follow trajectories with a balloon and measure constituents from it.

A NETWORK OF MEASURING STATIONS ON THE GROUND

The need for continuous measurements with reliable calibrations has led NASA to initiate the design of a new network of ground stations for monitoring stratospheric constituents. This "Early Detection System" is particularly aimed at measurements of ozone at 35 to 50 km (above its peak at 25 km). This is the altitude region where any changes in ozone are predicted to occur soonest. Dobson spectrophotometers measure the total overhead column, so that changes in ozone at, say, 40 km are masked by the variability in the much larger amount of ozone at 25 km.

The primary instruments in this network will be lidars to sense ozone and temperature. Such lidars have been developed, particularly at the French station in Haute Provence, and they have already routinely measured ozone and temperature at night, with a vertical resolution of better than 1 km up to 40 km, and further improvements in the power and sensitivity of these lidars are planned in the near future. Other instruments which would be deployed are:

(a) Dobson or Brewer spectrophotometers measuring column ozone, to ensure that the calibration of the lidars is consistent with that of the existing network. The same instruments, used in Umkehr mode, give the vertical distribution of ozone with a resolution of about 7 km.

(b) Visible spectrometers to measure the overhead column of NO_2 from scattered sunlight, measuring either selected parts of the spectrum (the Brewer spectrophotometer) or the whole spectrum (as pioneered by Noxon).

(c) Infra-red spectrometers or interferometers of very high spectral resolution, looking at the sun to measure the total overhead column of stratospheric HCl, HF, O_3, HNO_3, NO_2 and NO. A spectral resolution of 0.1 cm^{-1} allows one to distinguish the tropospheric from the stratospheric part of the column; higher resolution is important for other constituents with lower concentrations or weaker absorption lines, or lines masked by other absorbers (eg $ClONO_2$).

(d) Microwave sensors to measure emission from stratospheric O_3, ClO and H_2O_2 (and other constituents if low-noise detectors are developed at higher frequencies). Microwave measurements have the important advantage over infra-red measurements from the ground in that they give height information with a vertical resolution of about 7 km; but the bandwidth of typical microwave receivers is such that they give no useful measurements below about 25 km.

(e) Other sensors that may be developed (for example, the Zeeman Modulator Radiometer for measuring the overhead column of NO, and infra-red laser heterodyne spectrometers for sensing the stratospheric amounts of many constituents).

The measurement philosophy and structure of this network is still being debated. A modest number of sites is planned, probably 6 to 8. Important criteria for the lidars is clear skies at night, at perhaps 2000 m altitude to be free of scattering by tropospheric aerosol. The infra-red spectrometers require dry sites, preferably also high to reduce interfering lines from tropospheric CH_4 and N_2O. In some cases these spectrometers may be mounted at high mountain observatories some distance from the lidars and other instruments. One site will almost certainly be in USA, probably at Southern California's Table Mountain; another may be at Haute Provence in Southern France; other probable sites include Antarctica and Hawaii.

Gas	Probable range of volume mixing ratio (ppbv) between 20 and 25 km	Status of measurements of profiles
O_3	0.1 to 100 ppmv	*see chapter 6
Sources		
H_2O	2 500 to 8 000	well-measured, small differences between techniques
CH_4	500 to 1 600	*well-measured
N_2O	20 to 300	*well measured
CFCs	0.001 to 0.2	well-measured by samplers, balloons only
Long-lived reservoirs and sinks		
HNO_3	0.2 to 15	*well-measured
HCl	0.3 to 2	*well-measured from balloons, not from satellites
HF	0.05 to 0.5	measured from balloons, not from satellites
Radicals		
O	0.05 to 3	few measurements, one technique, balloons only
OH	0.01 to 1	*few measurements, large differences between the few techniques, balloons only
HO_2	0.05 to 0.5	few measurements, large differences between the few techniques
NO	0.2 to 20	*well-measured but not from satellites, large differences between techniques
NO_2	0.2 to 20	*well-measured, large differences between techniques
NO_3	0.001 to 0.3	*few measurements, large differences between the few techniques, balloons only
Cl	0.0001 to 0.02	few measurements, one technique, balloons only
ClO	0.005 to 2	well-measured from balloons and ground, large differences between techniques
Temporary reservoirs		
H_2O_2	0.01 to 2	few measurements, large differences between the few techniques, balloons only
N_2O_5	0.01 to 2	no profiles, few techniques
HNO_4	0.05 to 0.5	not definitive, one technique
$ClONO_2$	0.05 to 2	no profiles, one technique
HOCl	?	no measurements

Table 7.1 Minor constituents which interact with stratospheric ozone, and the status of measurements of their profiles. Those constituents whose vertical column in the stratosphere has been measured from the ground are marked *. NO and CFCs have also been measured from the Space Shuttle, but not with the global coverage associated with most satellites: like balloon measurements, they are not representative of the whole globe.

MODELS: DESCRIPTIONS AND COMPARISON WITH MEASUREMENTS 8

MODEL DESCRIPTIONS

The interaction of dynamical, radiative and photochemical processes in determining the structure of the stratosphere and of stratospheric ozone has been discussed in Chapter 5. The problem is three-dimensional in domain and a wide spectrum of spatial scales, ranging from gravity to planetary waves, is important. In attempting to describe the present state of the atmosphere and to predict its future development, numerical models, which can be integrated on high speed computers, are necessary. These models must themselves contain descriptions of the relevant processes and their interactions.

The basis of all photochemical models is the continuity equation or equation of mass conservation. The continuity equation describes the rate of change with time of a chemical constituent arising from chemical reactions and transport by atmospheric motions. Accordingly it is necessary to know (either from observations or by modelling) the three-dimensional motion fields and the relevant photochemical processes. A comprehensive stratospheric photochemical model would contain more than 30 constituents and approaching 200 reactions and, as stated above, atmospheric motions are variable on all time and space scales. It is for these reasons, and the concomitant high demands on computing resources, that several different kinds of simplified model have been developed.

Approximations are introduced into the models which aim to reduce both the number of constituent continuity equations which must be integrated and the number of spatial dimensions. For example, the very rapid photochemical equilibria which can be established between certain pairs of radicals means that only one continuity equation needs to be solved. The relative distributions are then computed using an equilibrium expression. Thus, for example, O_3 and O are in rapid photochemical equilibrium. Solution of the continuity equation is required for the sum $O + O_3$ and the equilibrium expression is then used to determine the partitioning between O and O_3. Similar treatments are used in most numerical models and the number of continuity equations can be reduced to around ten.

A further approximation involves reducing the dimensionality of the problem. Thus the continuity equation, appropriately averaged around a latitude circle, becomes the basis for a two-dimensional model with latitude and height as the variables. Alternatively, averaging over both latitude and longitude leads to a one-dimensional (height only) model of the atmosphere. In either case the averaged continuity equation contains terms for which there is no *a priori* description within the model. These terms, which contain the local departures from the average quantities, are referred to as eddy fluxes. Finally, box models, where transport into or out of the box is ignored, can be used. These models are useful for testing photochemical hypotheses. Recently they have been used successfully for studying the chemistry of air parcels moving along calculated trajectories.

The treatment of the eddy terms is the central problem in one- and two-dimensional modelling. In both types of model the usual procedure is to relate the eddy transport term to the mean gradient of the concentration using emperical transport coefficients. The theoretical justification for this procedure in one-dimensional modelling is poor. Nevertheless, one-dimensional models are capable of reproducing the gross features of the observed vertical distributions of many atmospheric constituents and their computational simplicity has made them very useful for investigating photochemical problems.

As discussed in Chapter 5, there has recently been considerable research activity in the field of two-dimensional tracer transport and significant advances have been made which have placed two-dimensional modelling on a firmer theoretical footing than hitherto. The relationships between transport by the mean motions and the eddies is now well understood as is the dependence of the transport on departures from radiative equilibrium. This departure itself also depends on eddy processes and this appears ultimately to limit the role of two-dimensional models: a completely interactive treatment is not available at present (see Chapter 5). Nevertheless treatments, which are physically well-based, for both mean and eddy transports are now available and a number of two-dimensional models has been developed. One type of two-dimensional model, for example, uses the Brewer-Dobson circulation shown in Figure 5.5. These two-dimensional models have been used to study both the present atmosphere and hypothetical perturbed atmospheres of the future.

For studies of the present day atmosphere and for predictions about its future development there are a hierarchy of models available for use, as discussed above. These models have their particular strengths and weaknesses. It is important that their limitations for any particular task should be understood.

One-dimensional models are extremely useful tools for studying sensitivities to changes in kinetic parameters. On the other hand their treatment of transport is poor. Assessment studies using one-dimensional models should best be regarded as indicative rather than quantitative.

Two-dimensional models contain very detailed chemistry as well as a reasonable treatment of atmospheric transport To this extent their predictions of future atmospheric states should be more useful than those by one-dimensional models. Nevertheless the treatment of transport in two-dimensional models is not self-consistent and thus results must still be treated with caution.

Three-dimensional models offer the best hope of internally

consistent dynamical photochemical calculations although such calculations are only just being attempted. However it should also be remembered that three-dimensional models themselves are not entirely free from empiricism. Furthermore, it will probably be difficult to diagnose the causes of the disagreements with observed ozone (for example) which will inevitably arise.

A good indication of the faith which can be placed in model predictions of the future is given by comparing model performance with observations. This is carried out in the following paragraphs. Assessment studies using these models are discussed in Chapter 9.

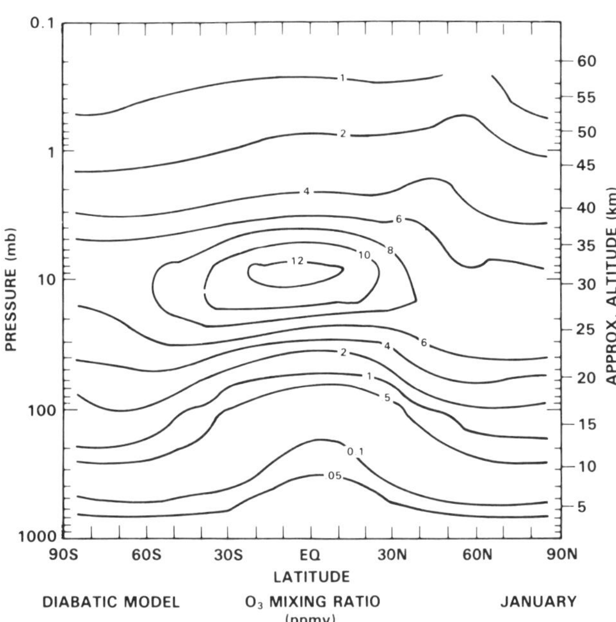

Figure 8.1a Ozone mixing ratio (ppbv) cross-sections, solstice conditions, calculated using the AER diabatic model

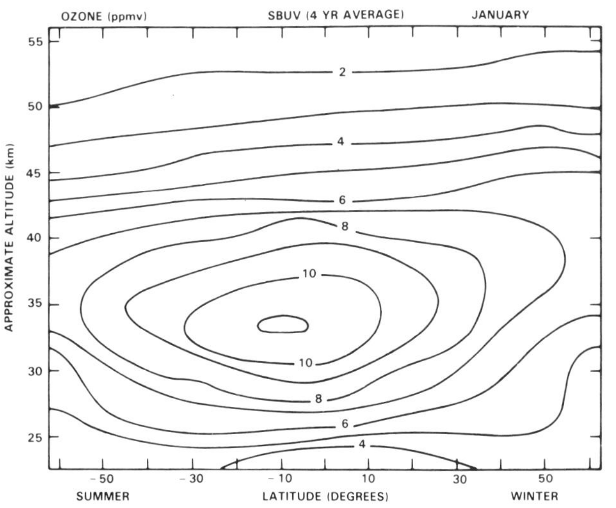

Figure 8.1b Measured two dimensional distribution of ozone mixing ratio (ppbv); a four year January average from SBUV data

COMPARISON WITH MEASUREMENTS

Comparing observed distributions of atmospheric constituents with those modelled generally shows reasonable agreement in terms of the main features of the spatial and temporal distributions. For example, Figure 8.1a shows a calculated latitude vs height cross section of ozone mixing ratio compared with satellite observations (Figure 8.1b). The main features are similar with, for example, peak mixing ratios found just above 30 km in low latitudes. At higher latitudes notice the somewhat larger mixing ratios found in both predictions and observations in the winter hemisphere compared with the summer hemisphere. This feature arises from the chemically driven inverse correlation between ozone and temperature. Figures 8.2a and 8.2b compare predicted and observed column ozone as a function of latitude and time. Again good agreement is seen

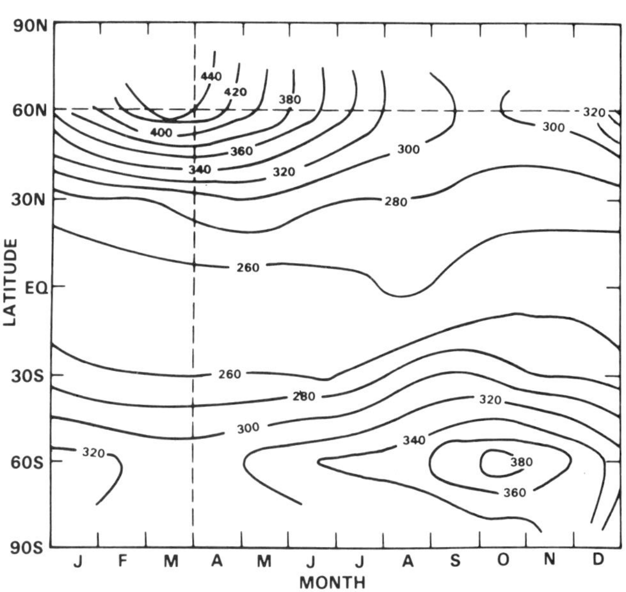

Figure 8.2a Column ozone values as a function of latitude and time of year: observations

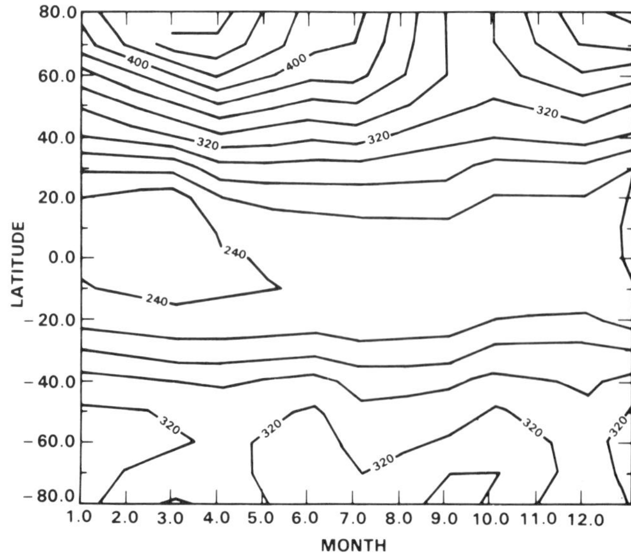

Figure 8.2b Column ozone values as a function of latitude and time of year: University of Oslo model calculations

with an equatorial minimum and higher column amounts found in the northern polar spring, a feature due to large scale transport.

The development of theoretical models has responded to the vast amount of satellite data which has become available recently. For example, Figures 8.3a and 8.3b compare satellite measurements of HNO_3 and a two-dimensional model calculation. The main features are modelled well, but there is a suggestion that our understanding of HNO_3 chemistry may be incomplete during winter in high latitudes. An additional source of HNO_3 during the high latitude winter, not included in the model, could be a possible explanation of the difference.

Figure 8.4a shows the zonal mean cross-section of N_2O for April measured from a satellite. An interesting feature is that, on a constant pressure surface, maximum mixing ratios are found in the sub tropics with a local minimum at the equator. This was unexpected prior to the acquisition of the satellite data; two-dimensional models predicted the maximum mixing ratios at the equator (and indeed this is observed at the solstice). It is now clear that this feature is related to an equatorial dynamical phenomenon, the semi-annual oscillation, which is driven partly by atmospheric wave motions generated in the troposphere. The effect of these waves can be included empirically in the models and good agreement between observed and modelled N_2O fields (Figure 8.4b) is then found.

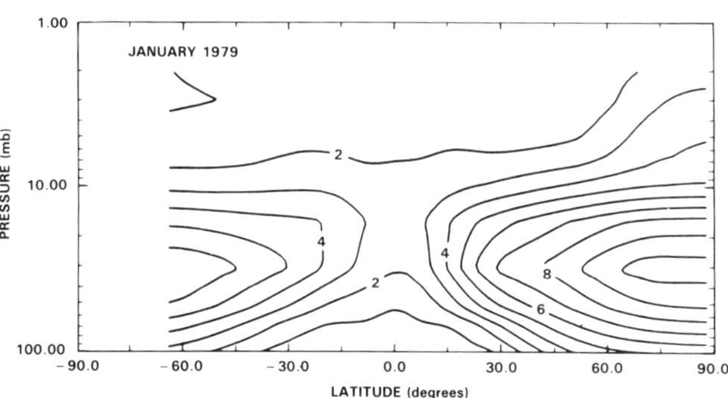

Figure 8.3a Monthly averaged zonal mean cross-sections of HNO_3 mixing ratio (ppbv) for January 1979

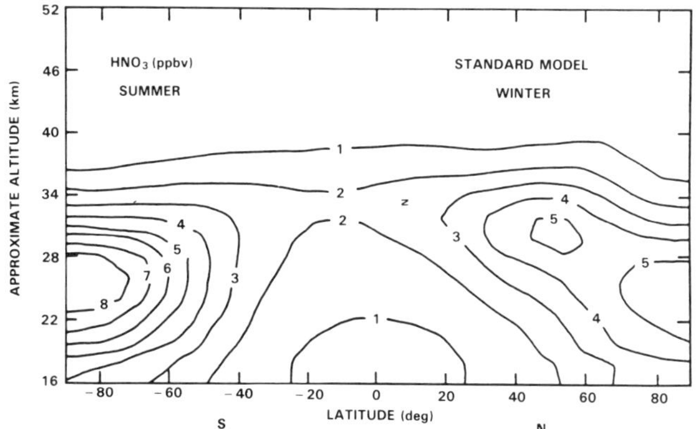

Figure 8.3b Calculations of HNO_3 distributions near winter solstice

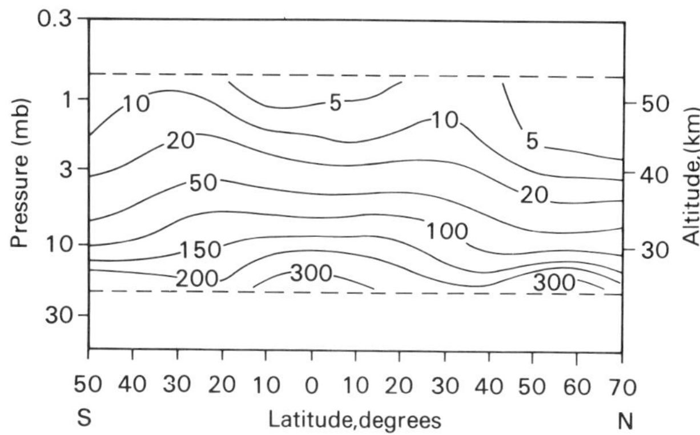

Figure 8.4a Monthly, zonal-mean cross-sections of nitrous oxide (ppbv) for April 1979 derived from SAMS observations

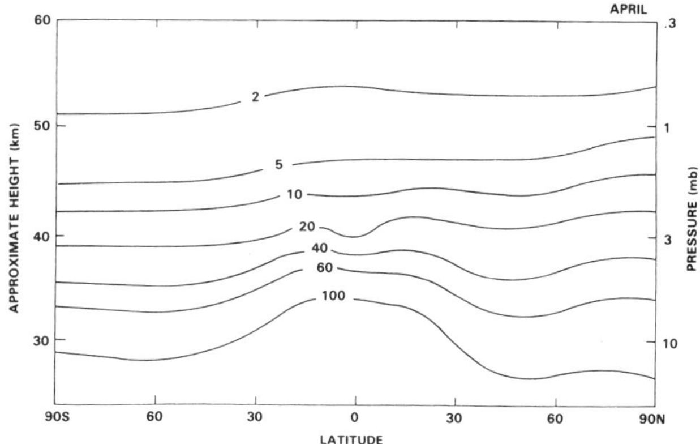

Figure 8.4b Nitrous oxide for April calculated from the Rutherford Appleton Laboratory/Cambridge University model

N_2O has a photochemical lifetime of months in the sunlight mid-latitude, middle stratosphere. In contrast the lifetime of NO_2 is on the order of about 100 seconds in that region. Comparison of observed and predicted NO_2 is thus a test of rapid photochemistry. Figure 8.5 shows balloon observations of NO_2 compared with model predictions. The principal features, for example the rapid changes at sunrise and sunset and the slow decay during the night appear to agree well.

A conclusion of the above is that numerical models can now reproduce many of the observed features of stratospheric tracer distributions. These distributions are due to both photochemical and transport processes; it is tempting to conclude that the current generation of models contain adequate treatments of these processes. However we will show below that a more detailed comparison of models and observations reveals a number of significant discrepancies. It is not clear what the ramification of these discrepancies is for the models' predictive capacity. Nevertheless, they must reduce our confidence in the use of models to predict future states of the ozone layer.

Figure 8.6 shows typical mid latitude observations of the ozone profile. Also shown are a range of model calcula-

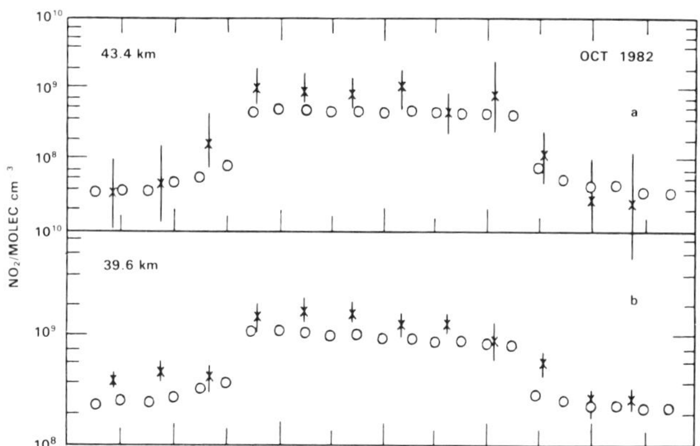

Figure 8.5 Diurnal variations of NO_2 concentration at 32°N, October, at (a) 43 km and (b) 40 km. * are observations by the Oxford pressure modulator radiometer; o represents model calculations

tions. The models underestimate the observed ozone by between 20 and 80% in the upper stratosphere, a region where the ozone distribution is controlled mainly by photochemistry. The discrepancy is unlikely to be due to inadequacies of transport although it could partly arise from this. Other causes could include inaccuracies or inadequacies in the representation of photochemistry, and errors introduced into the models by spatial averaging.

Figure 8.7 shows representative model profiles of the mixing ratio of active nitrogen compounds (NOy) in the stratosphere. The absolute concentration of NOy is important because interaction with chlorine compounds ameliorates the effect on ozone of increasing levels of Cl and ClO. Different models predict a wide range of NOy

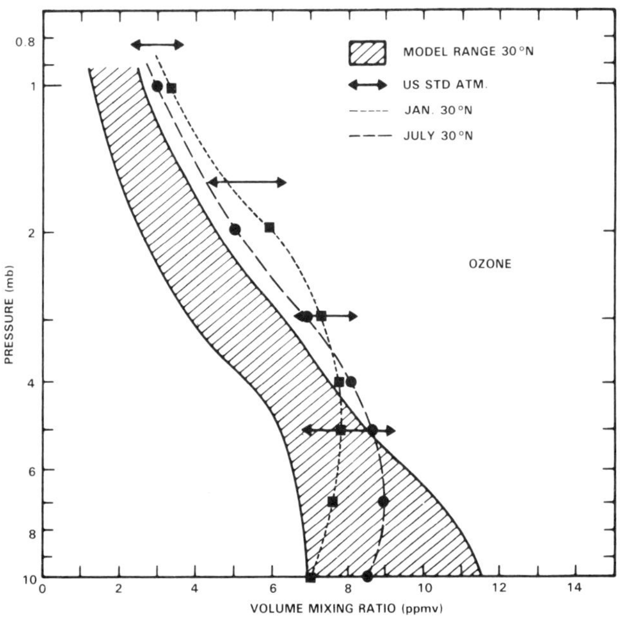

Figure 8.6 Vertical distribution of the ozone mixing ratio in the upper stratosphere. The shaded area includes 2-D model results for winter and summer at 30°N. Representative observations at 30°N for January and July are shown

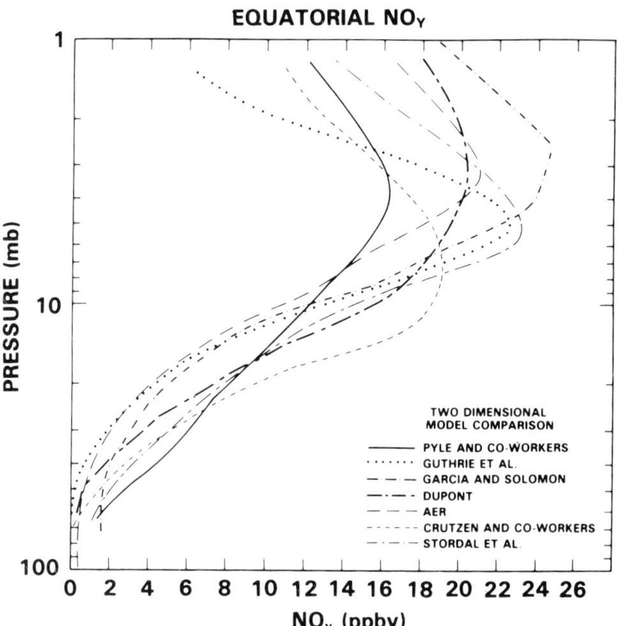

Figure 8.7 Two-dimensional model profiles of NO_y at the equator

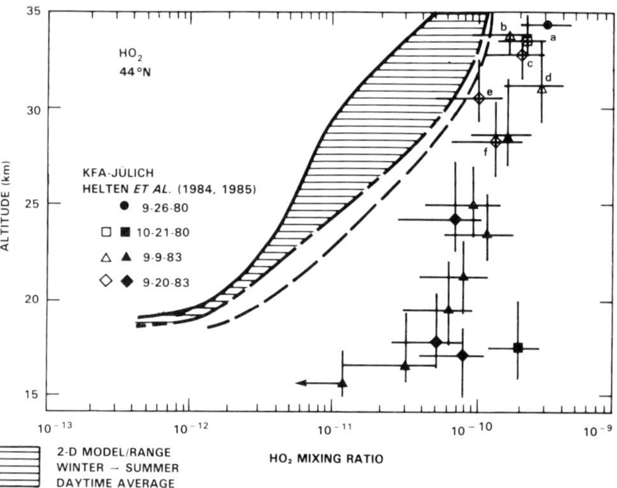

Figure 8.8 In situ observations of HO_2 by balloon. The dashed line is a typical 1-D model calculation for midday

concentrations both in the lower stratosphere and at the mixing ratio peak around 40 km. It is essential to resolve this discrepancy to improve confidence in the details of predictions of hypothetical stratospheres with high levels of Cl and ClO.

A third area of concern is the large discrepancy between modelled and observed hydrogen oxides, OH, HO_2 and H_2O_2. Recent measurements of HO_2 show larger mixing ratios than predicted, particularly in the lower stratosphere (Figure 8.8), while for H_2O_2 the predicted mixing ratios are larger than the observed upper limits (Figure 8.9). In all cases, more measurements are required, as well as a critical examination of the photochemical inputs to models.

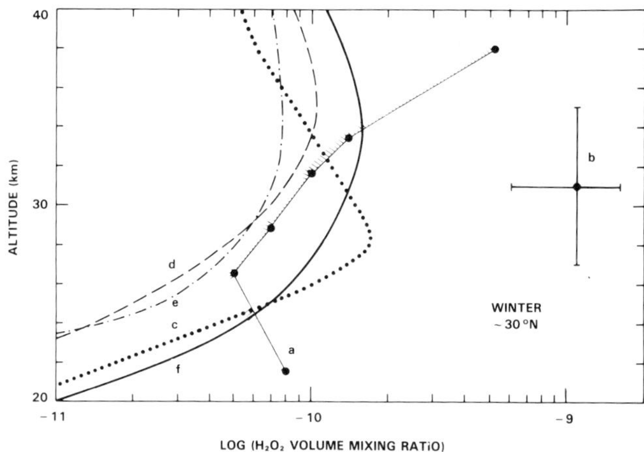

Figure 8.9 Observed and computed H_2O_2 profiles in winter at about 30°N. (a) is an upper limit using balloon borne far IR spectroscopy and (b) is a tentative detection from balloon-borne microwave limb sounding spectroscopy. (c), (d), (e) and (f) are 2-D model calculations

Other discrepancies also exist. For example, observed HNO_3 is generally low compared with modelled values above 30 km; models generally appear to be unable to reproduce the observed profiles of both N_2O and CH_4 with the same transport parametrization. These, and the problems discussed in the above paragraphs are all areas of active research. Solutions must be found in order to assess our confidence in the model predictions reviewed in Chapter 9.

Because of the uncertainty associated with averaging in 1-D and 2-D models, a Lagrangian approach has recently been formulated. In this approach the trajectory of an arbitrarily defined parcel of air is traced using meteorological observations. If the air parcel remains intact then its chemical evolution can be studied using a photochemical box model, with the location of the box determined by the trajectory. No mixing between the box and the surroundings if allowed in the calculation and the meteorological and photochemical problems are thereby separated. Experience suggests that this separation is a reasonable approximation up to about ten days. If observations are available comparison between theory and measurements is possible. For example, if an air parcel is arbitrarily defined at the location of the LIMS satellite observation of O_3, NO_2, HNO_3 and H_2O, and SAMS observations of CH_4 and N_4O, it turns out in practice that during a 10-day period there are between about 15 and 25 further observations along the trajectory of the air parcel. In principle there is thus a framework for testing the photochemical evolution of the real atmosphere at a specific time and in a specific place, with the spatial averaging reduced to a scale of about 5 great circle degrees (about 560 km) and 5 km in the vertical, and with little or no temporal averaging. The method is limited at present by the fact that observations of only 6 of the 32 constituents calculated are available, and that the observational errors are too large to make 15-25 individual observations sufficient. Nevertheless useful tests of theory can be carried out. For example, discrepancies between theory and observations have been shown to exist at some altitudes for ozone and for nitric acid, while good agreement was found for the day-night difference of NO_2. Figure 8.10 shows an example.

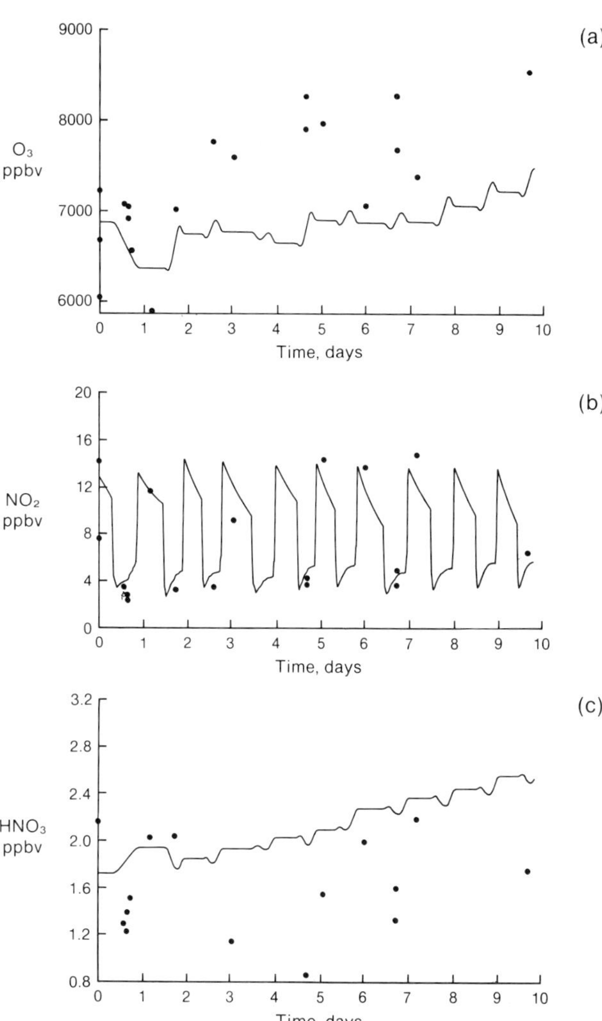

Figure 8.10 ● LIMS observations of ozone, NO_2 and HNO_3 along a trajectory at 35 km, over ten days in February 1979. The curve is a calculation by Lagrangian model. LIMS observations are within a 5° radius of the model trajectory

The comparisons between models and measurements discussed here have largely dealt with average vertical profiles, and average zonal mean cross-sections. Such comparisons have been very informative about the major features of the large scale transport, and have also indicated rather subtler problems, for example the existence of a nitric acid source in mid stratosphere during the arctic winter Nevertheless, such comparisons do not constitute a sufficiently severe test of the model calculations, because of uncertainties caused by the averaging in 1-D and 2-D models, and also because of the errors associated with the satellite data. A current research objective is the construction of 3-D models including detailed photochemistry; parallel developments, avoiding many of the problems of atmospheric transport, are in the area of trajectory studies, discussed above.

CONFIDENCE IN MODELS

We have stated that models can now reproduce many observed features of the distribution of trace gases in the stratosphere. On the other hand, comparisons of some specific features reveal inadequacies in our present understanding which limit our confidence in the use of models for prediction purposes. We cannot quantify the reduction in confidence.

The magnitude of any calculated ozone depletion in the future depends on the concentrations of radicals which destroy ozone, relative to the reservoir species. Few measurements exist for many of the reservoirs, so that the validation of these vital aspects of stratospheric chemistry cannot be carried out adequately.

MODEL PREDICTIONS OF OZONE CHANGE 9

INTRODUCTION

The stratosphere has been modelled in increasing detail in order to estimate the effects of human activity on the ozone layer. It is important that we should understand how the atmosphere responds to a variety of changing conditions which may already have occurred or may occur in the future. The amount of ozone in the atmosphere is affected by emissions of several source gases, in particular chlorofluorocarbons (CFCs), methane (CH_4), nitrous oxide (N_2O) and carbon dioxide (CO_2). The CFCs are transported up into the stratosphere where they are mainly destroyed by sunlight to produce the reactive chlorine which can take part in ozone-destroying cycles. Among the threats to stratospheric ozone two CFCs, $CFCl_3$ (CFC 11) and CF_2Cl_2 (CFC 12), have remained the principal ones because of their inertness in the troposphere and their continued emission from industrial processes. The lifetimes of $CFCl_3$ and CF_2Cl_2, for example, are approximately 65 and 111 years respectively which means that any effect they have will be long-lived and there will be a substantial time-lag between increases (or reductions) in CFC release and associated changes to the abundance of ozone in the atmosphere.

Since we cannot perform experiments with the real atmosphere the only way to predict changes to the ozone layer is to use a computer model. The first step in a modelling study is to consider one individual perturbation (for example, a specified increase in CFC release) while all other aspects of the model are held constant. The ozone amount is then compared with a similar model calculation that did not contain the perturbing influence. A more realistic experiment is one in which the perturbation is time-dependent, for example an increase in CFCs of a certain percent each year. Next, more complex calculations may be devised in which the concentrations of the entire suite of important gases are varied, either in steady-state or time-varying calculations. These calculations may include temperature feedback effects. Examples of all of these types of calculations are presented later. Finally, the radiative effects of all the important gases, including CO_2, should be included in a coupled chemical-dynamical calculation, although this has not been done to date.

One-dimensional models (with height as the variable coordinate) have been extensively employed in perturbation studies because they are computationally fast and calculations may be continued for many simulated years. Recently, there has been an increased emphasis on two-dimensional modelling (with height and latitude as the variables) due to the realization of the importance of latitudinal and seasonal effects, increased computing facilities and the availability of satellite data to test these models.

Model predictions are usually presented as changes in the ozone column and/or the local change in ozone as a function of height. The column of ozone is the total amount of ozone above a unit area at the earth's surface and so it is important in determining changes to the total UV radiation which reaches the earth's surface. On the other hand, height profiles are more useful in giving insight into the various mechanisms which affect the ozone layer; it is often the case that large increases and reductions of ozone at different heights may cancel to give a small change in column ozone.

One-dimensional model predictions of ozone column changes due to increases in CFCs have changed widely during the last 10 years, from high values of up to 20% to the more recent lower values which are discussed in this text. These changes are largely due to the revision of chemical reaction rates and to the inclusion of reactions which had been overlooked earlier.

SCENARIO STUDIES

A comprehensive study was carried out by a number of modelling groups as part of the recent Assessment (NASA/WMO, 1986). Each of the computer models had similar chemistry (as given in Appendix A to the Assessment), but other processes such as transport mechanisms, diurnal averaging and photodissociation cross-sections were treated differently. A series of hypothetical situations, or 'scenarios', were investigated by both one and two-dimensional models in steady-state and time-dependent calculations.

ONE-DIMENSIONAL MODELS

Steady-state model runs

The use of steady-state model runs allows a simple examination of a perturbing influence on the atmosphere. The initial state is specified by fixing either the concentration of the perturbing species or its flux from the ground to the lowest model level. The model is then run for many simulated years with the concentration or flux held constant at the perturbed value until the modelled atmosphere has adjusted to the imposed perturbation and a 'steady-state' is achieved. Model results are taken from this final state; the complete history of the run need not be described. A steady-state model run cannot refer to a specific time in the future but it may be chosen to have conditions typical of a future time.

Increases in chlorine

Figure 9.1 illustrates the changes to the steady-state ozone height profile as a result of increasing the total chlorine in a one-dimensional model to 8 ppbv while all other species were kept constant. The associated column ozone depletion was 5.7%. When the amount of ozone is changed in the

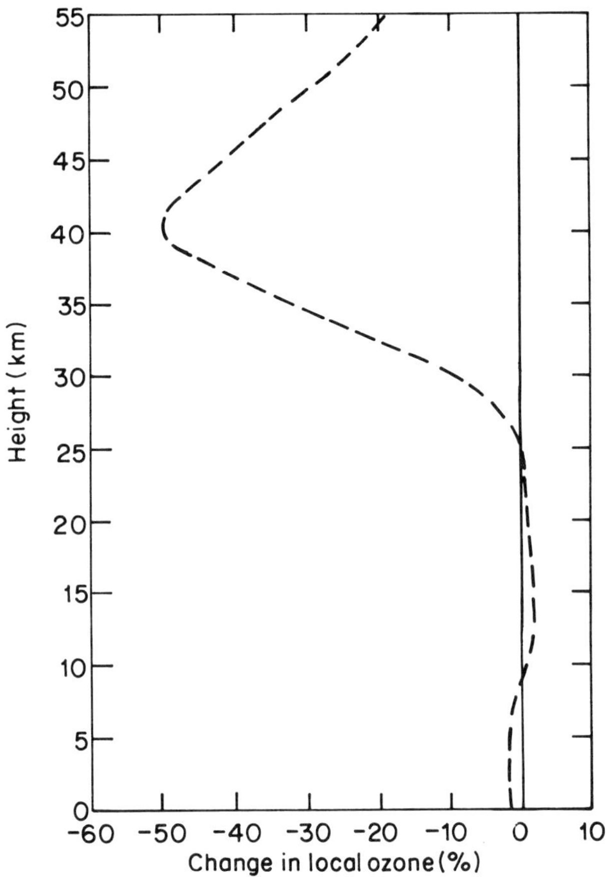

Figure 9.1 Height profile of the percentage change in ozone, predicted by Model A (with temperature feedback) at steady state, due to an increase in stratospheric chlorine from 1.3 ppbv to 8 ppbv. The predicted change in column ozone was – 5.7%. (These results, and those shown in Figures 9.2 to 9.5, are from the Lawrence Livermore National Laboratory 1-D model, and are reproduced with the permission of Dr D Wuebbles.)

model due to the increase in chlorine, the temperature also changes and this in turn will feed back on to the ozone chemistry. This is known as "temperature feedback" and is taken into account in the model. 8 ppbv of chlorine is approximately the amount of chlorine which would result if CFC emissions continued at the present rates long enough for a steady state to obtain. If the rates were doubled then approximately 15 ppbv would result. Column ozone changes for these two scenarios predicted by a variety of one-dimensional models are shown in Table 9.1. The predictions are relative to a model run with about 1.3 ppbv background chlorine. The wide variation in the model predictions indicates an unexpected sensitivity to chlorine in these models. One of the major factors in these discrepancies is the different amounts of reactive nitrogen in the models. The nitrogen species are important both because of their role in interfering with the chlorine catalytic cycle (see Chapter 4) and their direct role in catalysing ozone destruction. Resolution of this major discrepancy is required to put the model predictions on firmer ground.

Table 9.1

	Percentage column ozone change predicted by model: A	B	C	D	E
8 ppbv Clx	-5.1 (-5.7)	-2.9	-4.6	 (-4.1)	 (-9.1)
15 ppbv Clx	-12.2 (-12.4)	-17.8	-15.0	 (-8.8)	 (-22.0)

Percentage column ozone changes predicted by a variety of one-dimensional models at steady-state, due to increases in chlorine from (a) 1.3 ppbv to 8 ppbv, and (b) from 1.3 ppbv to 15 ppbv. Numbers in brackets refer to model runs with temperature feedback.

Nonlinearity

Until recently it was thought that the percentage change in column ozone relative to the amount of chlorine in the stratosphere was approximately constant. However, computer predictions have shown some nonlinearity: once the amount of chlorine exceeds a certain threshold, then each additional chlorine atom will destroy progressively more and more ozone. In this way the ozone depletion will take place increasingly quickly, even at constant emission rates, until eventually a rapid depletion will occur. Table 9.1 illustrates this point well: although the increase in chlorine to 15 ppbv is approximately double the increase to 8 ppbv, the models do not predict a simple doubling of the ozone depletion. Model B, for example, is extremely non-linear; at 8 ppbv it predicts a depletion of 2.9%, but at 15 ppbv it predicts a depletion that is about six times that amount (17.8%). The nonlinearity is strongly linked to the amount of odd nitrogen in the models. The process can be considered chemically as a 'titration' from nitrogen dominance to chlorine dominance of stratospheric chemistry with chlorine nitrate as the buffer. Resolution of the differences in odd nitrogen calculated in the models is obviously important. However, recent improved two-dimensional models have given predictions that are close to linearity. In particular, one recent study (Stordal and Isaksen 1986) predicts that the level of chlorine at which ozone depletion becomes markedly non-linear is much higher than at first thought (around 25 ppbv).

Increases in other source gases

The changes to the ozone profile predicted by model A as a result of scenarios involving other source gases – methane (CH_4), nitrous oxide (N_2O) and carbon dioxide (CO_2) – are shown in Figure 9.2; temperature feedback is included in each case. The associated column ozone changes are shown in Table 9.2. Increases in CFCs and N_2O both lead to a decrease in column ozone but increases in CH_4 and CO_2 both lead to an increase in column ozone. Predictions of column ozone change predicted by various other one-dimensional models are also given in Table 9.2.

Table 9.2

	Percentage column ozone change predicted by model: A	B	C	D	E	F
2 X CH_2	+2.0 (+2.9)	+3.0	+0.9	(+1.6)	(+1.4)	+1.7
1.2 X N_2O	−2.1 (−1.7)	−2.6	−1.8	(−1.1)	(−1.2)	−2.3
2 X CO_2	(+3.5)		(+2.6)	(+3.1)	(+1.2)	(+2.8)

Percentage column ozone changes predicted by a variety of one-dimensional models at steady-state, due to increases in gases other than CFCs. Numbers in brackets refer to model runs with temperature feedback.

Simultaneous increases in source gases

Results of combined or 'multiple-perturbation' one-dimensional scenarios in which several of the source gases are perturbed in the same model run are shown in Table 9.3. Note that the effects of these concurrent increases are not additive. For example, from Tables 9.1 and 9.2 each separate perturbation to model A predicts:—

	% ozone change
8 ppbv Clx	−5.7
2 X CO_2	+3.5
2 X CH_4	+2.9
1.2 X N_2O	−1.7

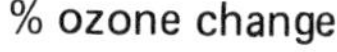

−1.0%

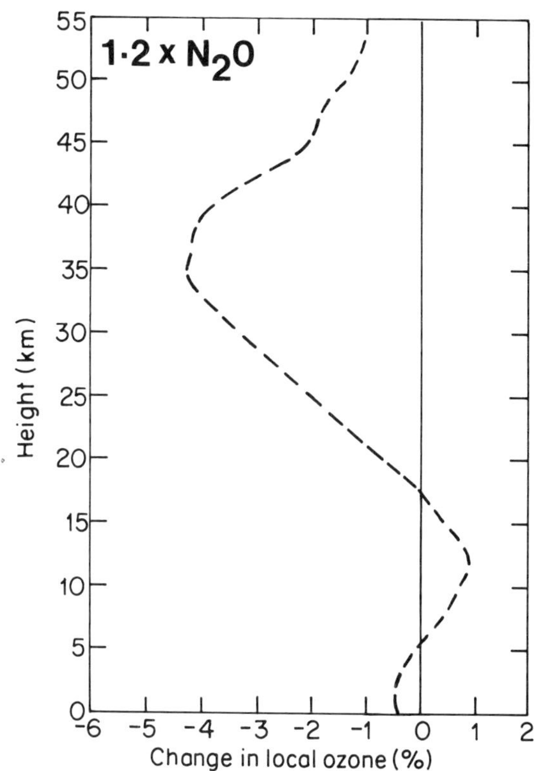

Figure 9.2b Height profile of the percentage change in ozone predicted by Model A (with temperature feedback) at steady state due to a 20% increase in nitrous oxide. Predicted change in column ozone: −1.5%

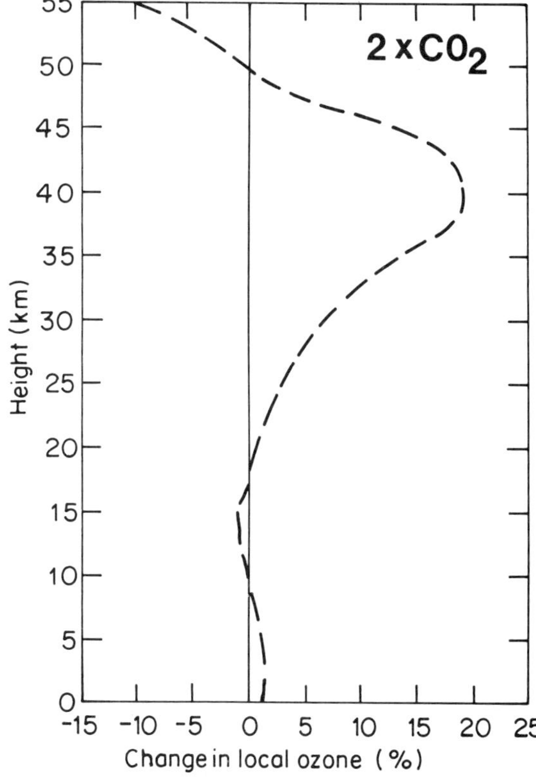

Figure 9.2c Height profile of the percentage change in ozone predicted by Model A (with temperature feedback) at steady state due to a doubling of carbon dioxide. Predicted change in column ozone: +3.5%

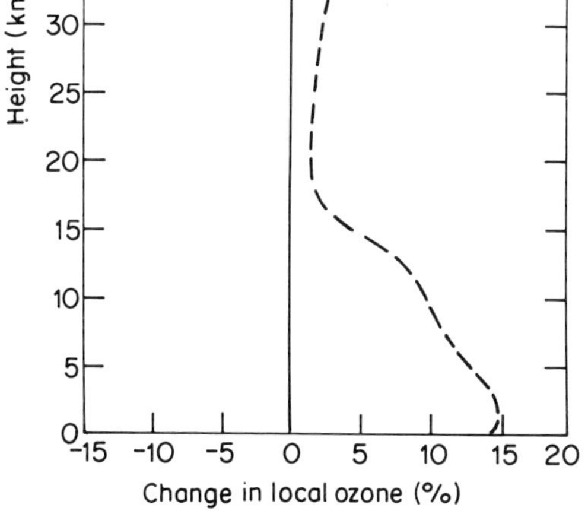

Figure 9.2a Height profile of the percentage change in ozone predicted by Model A (with temperature feedback) at steady state due to a doubling of methane. Predicted change in column ozone: +2.9%

However, when all of these perturbations were introduced together in the same model run the predicted change was +0.2% (Table 9.3). The effect is more pronounced for the 15 ppbv Clx case which gave −4.6% in the combined calculation while the individual components add up to −7.7%.

Table 9.3

	Percentage column ozone change predicted by model: A	B	C	D	E	F
8 ppbv Clx +2 X CH_4 +1.2 X N_2O	−3.4 (−2.8)	−3.0	−3.3	(−2.3)	(−6.0)	−3.1
8 ppbv Clx +2 X CH_4 +1.2 X N_2O +2 X CO_2	(+0.2)			(0.0)	(−5.2)	(−1.4)
15 ppbv Clx +2 X CH_4 +1.2 X N_2O	−7.8 (−7.2)	−8.2	−8.8	(−5.6)	(−13.7)	−7.2
15 ppbv Clx +2 X CH_4 +1.2 X N_2O +2 X CO_2	(−4.6)			(−3.5)	(−13.6)	

Percentage column ozone changes predicted by a variety of one-dimensional models at steady-state due to the concurrent increase of several gases. Numbers in brackets refer to model runs with temperature feedback.

Height distribution of ozone changes

To assess the absolute effect of various scenarios the height profiles of ozone change shown in Figures 9.1 and 9.2 must be interpreted in conjunction with the height profile of ozone shown in Figure 9.3. The peak in ozone concentration occurs at approximately 25 km so that quite small percentage changes at around this height have a significant effect on the total column of ozone. Conversely, percentage changes at other heights have a lesser effect. For example,

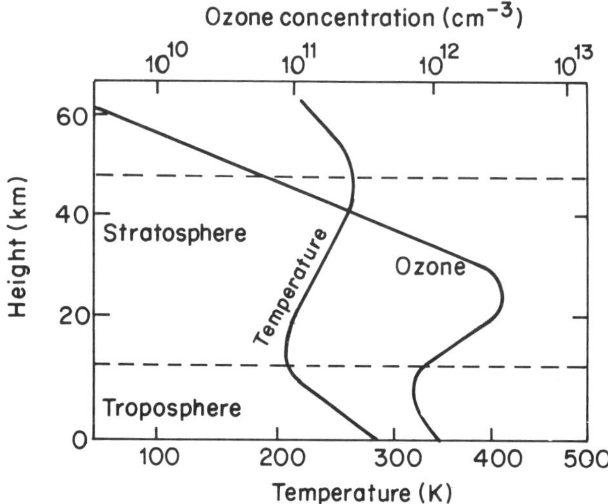

Figure 9.3 The height profiles of temperature and ozone concentration in the atmosphere

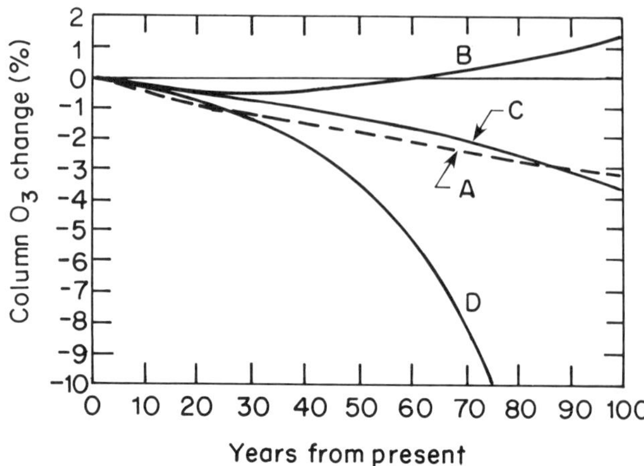

Figure 9.4 Changes in column ozone predicted by a 1-D model as a function of time with the following scenarios:

A: CFC emissions at 1980 levels, plus carbon dioxide emissions increasing at about 0.5% per year

B: as scenario A and with methane and nitrous oxide increasing at 1% per year and 0.25% per year respectively

C: as scenario B but with CFC emissions increasing at 1.5% per year

D: as scenario B but with CFC emissions increasing at 3% per year

Figure 9.1 shows a total column depletion of only 5.7% even though the ozone decrease at 40 km was approximately 50%. The predicted maximum decrease in ozone at about 40 km agrees well with the theoretical height profile of the chlorine catalytic cycle which peaks near 40 km. Increases are predicted in the lower stratosphere (ie 10-25 km) primarily due to the ozone recovery (or self-healing) mechanism (increased UV penetration, greater O_2 photolysis and resultant ozone production at lower altitudes as ozone is destroyed above) and also due to the chlorine interfering with the dominant catalytic cycles in the lower stratosphere.

Time-dependent model runs

Several one-dimensional models were employed to run time-dependent calculations, which are regarded as the most realistic of the one-dimensional model assessments. Figure 9.4 shows results from one of these models which was run from the present to 100 years in the future. All model runs were started with 1980 emission rates. Curve A shows ozone changes with CFC emission rates at 1980 levels and CO_2 increasing at approximately 0.5% per year. Curve B shows CFC release at 1980 levels but with CH_4 concentration increasing at 1% per year, N_2O at 0.25% per year in addition to the CO_2 increase. Comparison of the two illustrates the effects of simultaneous emissions of these other gases on the ozone depletion. However, curves C and D show results for the same scenario as curve B except that CFC release is increasing at 1.5% per year and 3% per year, respectively. In the former case, column ozone

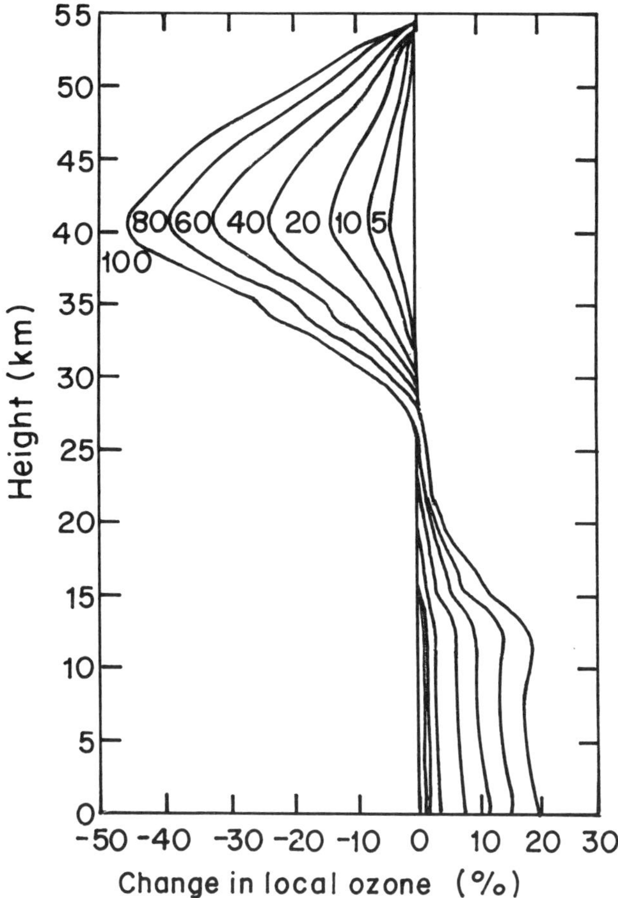

Figure 9.5 Height profiles of percentage ozone changes predicted by a 1-D model at selected times (5 to 100 years) for scenario C shown in Figure 9.4

depletion was calculated to be about 3% over the next 70 years, but with a CFC growth rate of 3% per year the predicted ozone depletion is about 10% after 70 years and rapidly accelerating. In the latter case the effects of the chlorine were initially masked by the other gases but eventually chlorine dominated the ozone loss. In Figure 9.5 the height distribution of ozone change in run C (1.5% per year increase in CFC plus simultaneous increases in other source gases) is shown; large percentage decreases in ozone are seen around 40 km with smaller percentage increases below 25 km.

TWO-DIMENSIONAL MODELS

Latitudinal variation of ozone change

The percentage change in steady-state ozone column content calculated by a two-dimensional model with 1980 CFC emission rates is shown in Figure 9.6, relative to an atmosphere with 1.3 ppbv chlorine. The steady-state chlorine in the perturbed model was 8.2 ppbv. A large latitudinal variation is evident, with much larger depletion at high latitudes than at low latitudes. As shown below (Figure 9.8) this arises because the self healing mechanism is less efficient in high latitudes. A significant, but smaller, seasonal variation

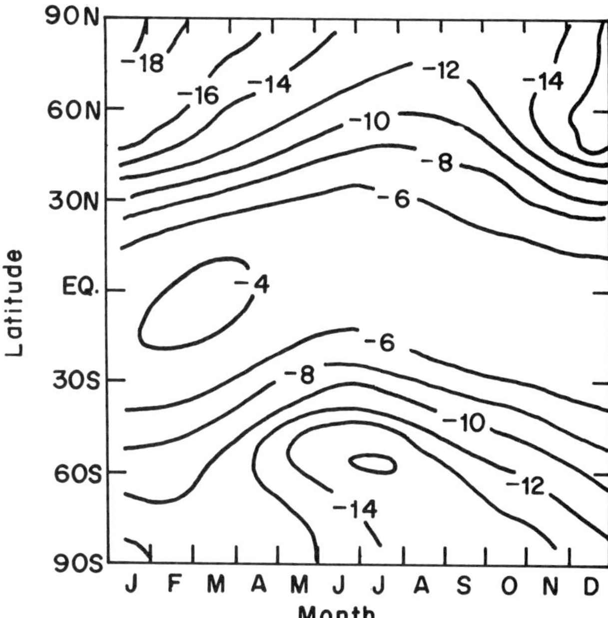

Figure 9.6 Change in column ozone predicted by a 2-D model as a function of latitude and season, in response to an increase in chlorine of approximately 7 ppbv

is also present. The globally-averaged column ozone depletion for this model was 8.5% which falls at the high end of the range predicted by the one-dimensional models. However, Figure 9.6 shows that the one-dimensional models severely underestimate the local ozone reductions of up to 18% that are predicted at high latitudes. There are some differences in the latitudinal gradient of the ozone depletion predicted by various models. Figure 9.7 shows the latitudinal variation of column ozone depletion for the month of April from two different models in response to an increase of chlorine of approximately 7 ppbv (but no other gases increasing). The ratio of the ozone depletion at the poles to that at the equator varies from 2:1 to 4:1 depending on the model. The differences between the two predictions are believed to stem from differences in the formulation of the dynamics in these models. Recent

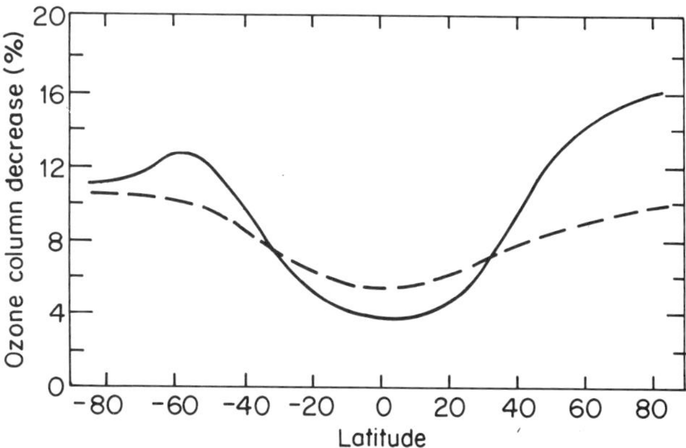

Figure 9.7 The latitudinal variation of the percentage decrease in column ozone predicted by two different 2-D models for the month of April, in response to an increase in chlorine of 7 ppbv

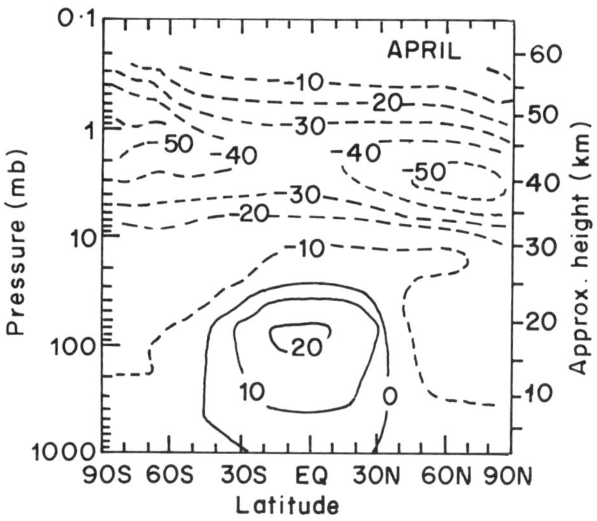

Figure 9.8 Latitude-height distribution of percentage ozone change for April predicted by a 2-D model in response to an increase in chlorine of 7 ppbv

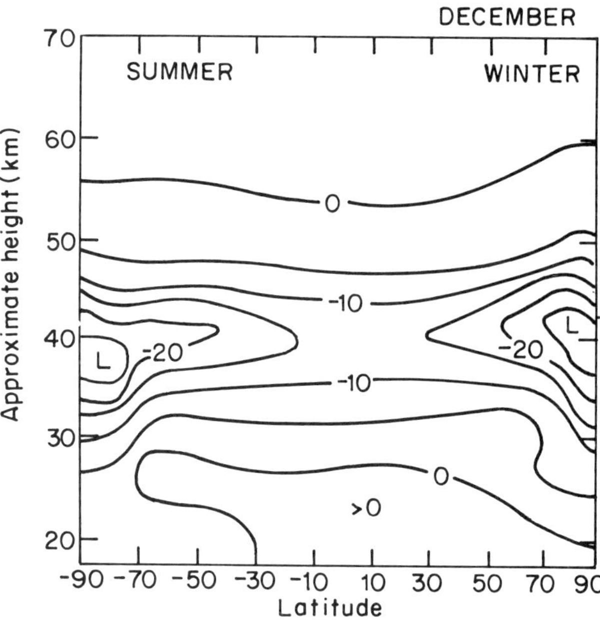

Figure 9.9 Percentage changes in ozone between 1965 (1.3 ppbv chlorine in the stratosphere) and 1985 (2.7 ppbv chlorine), predicted by a 2-D model

improvements in our theoretical understanding of these models and further testing of the models using satellite data should help to resolve these differences.

Height distribution of ozone change

The height distribution of percentage ozone change predicted by a two-dimensional model in April with 8 ppbv chlorine and other gases at present levels is shown in Figure 9.8. Large percentage decreases are predicted at around 40 km which is in agreement with one-dimensional models. Substantial latitudinal variation is evident in the lower stratosphere below 25 km. An increase in ozone is predicted at these levels in equatorial latitudes and a decrease in polar regions, with a switch-over occurring in mid-latitudes at approximately 40°. This low altitude variation of the ozone response is primarily responsible for the latitudinal structure in ozone column changes seen in Figure 9.6. Figure 9.8 emphasizes the inability of one-dimensional models to give an adequate description of the local response of ozone to CFCs, particularly at mid and high latitudes.

The percentage increase in column ozone at low altitudes in equatorial regions (Figure 9.8) is similar to that predicted by one-dimensional models (Figure 9.5) and is due to the photochemical self-healing effect described earlier. Below 25 km at middle and high latitudes, however the concentration of ozone is principally determined by dynamical factors. At these latitudes photochemical self-healing occurs too slowly, compared with these dynamical effects, to increase the ozone significantly at low altitudes and the perturbing influence of CFCs on ozone is therefore large. Other important factors determining the latitudinal gradient of ozone depletion predicted by the model include the strength of the horizontal mixing of the gases compared with their transport by the mean motion, a factor which is sensitive to the formulation of the model dynamics.

Predicted ozone changes 1965-1985

The total chlorine content of the atmosphere has increased from approximately 1.3 ppbv in 1965 to approximately 2.7 ppbv in 1985. Two-dimensional model calculations have been carried out which incorporate this increase in chlorine (but no concurrent increases in other gases). Figure 9.9 shows the calculated percentage ozone change for this period from a two-dimensional model. This change should be regarded as the upper limit of ozone changes predicted by the models, as the mitigating effects of simultaneous methane and carbon dioxide increases have not been included. The models predict the largest local ozone reductions at 40 km. However, because most of the ozone is located at low altitudes (see Figure 9.3) the column ozone reduction predicted by this model is small. For example, the predicted column ozone depletion in polar regions is only a few percent. This is much smaller than recent observations in the Antarctic which measured seasonally varying reductions of 5-40% during the period 1980-1984 relative to 1957-1973. This inability to model the large changes which are already being observed in the Antarctic is a matter for serious concern.

Time-dependent multiple perturbation calculations

A recent two-dimensional time-dependent model run (Stordal and Isaksen 1986) has included increases of 1% per year in CH_4 and 0.25% per year in N_2O in addition to CFC release. The model was run from 1960 to 2030. After 1980

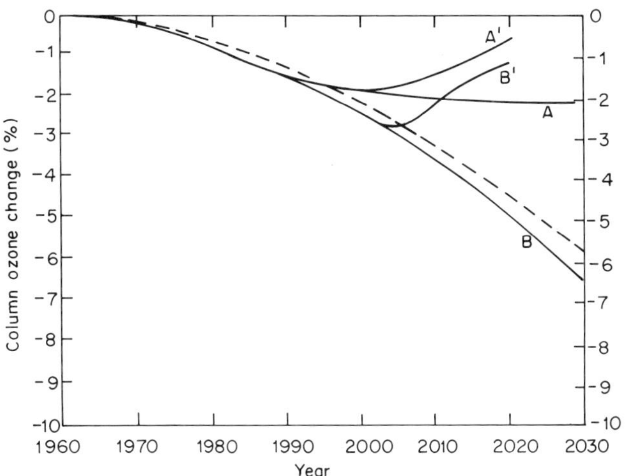

Figure 9.10 Percentage changes in globally-averaged column ozone with time predicted by various runs of a 2-D model. All model runs included increases in methane (1% per year) and nitrous oxide (0.25% per year). The effect of CFC emissions continuing at 1980 rates, and increasing at 3% per year are shown in curves A and B respectively. Curves A' and B' indicate column ozone changes predicted as a result of the complete cessation of CFC releases in the year 2000. (The results shown in Figure 9.10 to 9.12 are from the University of Oslo 2-D model, and are reproduced with the permission of Dr Ivar Isaksen)

several different scenarios were calculated: in run A the CFC release was kept constant at 1980 rates and in run B the release was increased at the rate of 3% per year. Figure 9.10 shows the evolution of the ozone change. For constant CFC emission the global ozone depletion reached approximately 2% in the year 2000 and did not increase further; the decrease in ozone due to CFC and N_2O increase was compensated by the increase in CH_4. In run B however, the CH_4 counterbalance was no longer sufficient and following a slow initial change the ozone depletion

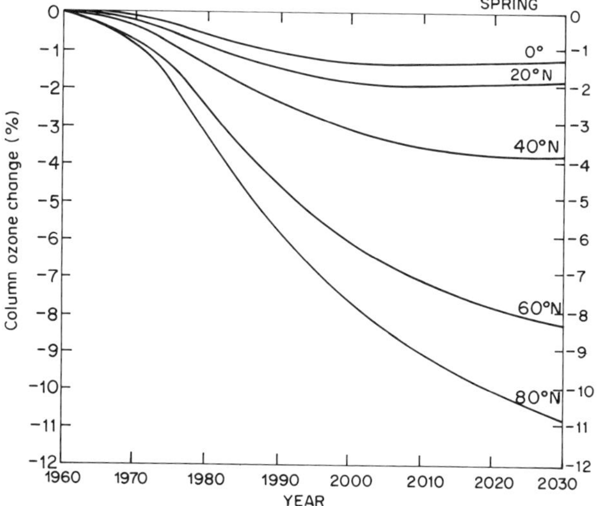

Figure 9.11 Percentage changes in column ozone at different latitudes predicted by a 2-D time dependent model run with CFC emissions at 1980 rates, nitrous oxide growing at 0.25% per year and methane at 1% per year

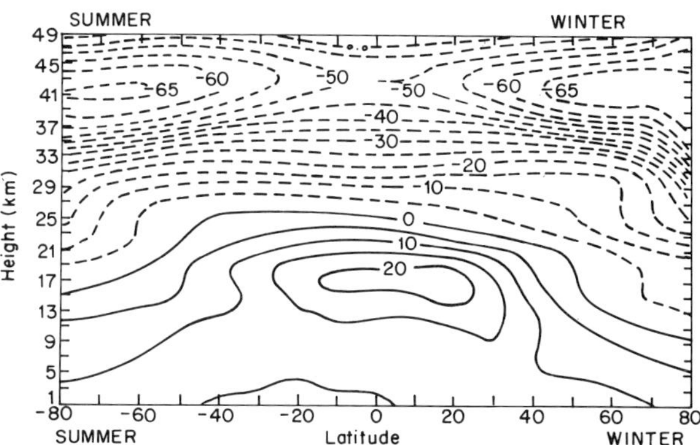

Figure 9.12 Latitude-height cross section of ozone change predicted for the year 2020 by a 2-D model run in which methane and nitrous oxide increased as in Figure 9.11 and CFC increased at 5% per year until 2000, and at 3% per year thereafter

exceeded 6% by 2030 and continued to grow. These results are qualitatively similar to those of the one-dimensional models shown in Figure 9.4. In a subsequent paper (Isaksen, 1987) the effect of increasing CO_2 and of temperature feedback were included in run B (dashed curve, Figure 9.10); although the depletion is reduced, the basic trend of the ozone depletion is unaffected. The latitudinal variation of ozone change in runs A and B shows considerably larger differences between equator and pole than the runs which considered CFC release only (Figures 9.6 and 9.7). In run A, for example, the ratio between the change in column ozone at high latitudes and the equator reached a factor of 8 in the year 2030 (see Figure 9.11). In a third run (run C), in addition to increases in CH_4 and N_2O (but with no temperature feedback) the CFC emissions were increased at about 5% per year until the year 2000 and 3% per year thereafter. The chlorine in the model reached approximately 9 ppbv by the year 2030. Figure 9.12 shows the calculated latitude-height cross-section for ozone changes in January 2020 of this run. The upper stratosphere ozone depletion was quite similar to the steady-state, CFC-release only runs shown in Figure 9.8. The largest percentage ozone depletion occurred at high latitudes near 40 km. A more pronounced ozone increase is indicated in the troposphere and lower stratosphere due to the effects of the methane increase.

Several observations may be drawn from these computer predictions. The choice of specific source gas scenarios can drastically affect the ozone changes predicted by these models. Little change in globally-averaged ozone may be expected in the next few decades unless a significant sustained growth in CFC emissions or drastic differences in present growth rates of other gases were to occur; however, beyond 20-30 years ozone decreases could be substantial.

Ozone recovery

Owing to the extremely long lifetimes of the CFCs there

has been concern over the extent to which the ozone will continue to deplete even after CFC emissions are reduced. In Figure 9.10 the branching curves marked A′ and B′ show the ozone changes following a complete cessation of CFC emissions in the year 2000. In both cases column ozone continues to decline for 5-10 years before a slow recovery begins but the additional depletion is less than 1 percent. The overshoot would naturally be greater if CFC emissions are reduced rather than terminated completely. These predictions depend critically on the assumptions of changes in other source gases, particularly methane.

MODEL UNCERTAINTIES

The value of model predictions of ozone change cannot be assessed without consideration of the sensitivity of the results to uncertainties in the model input data and assumptions. Recognized uncertainties include (a) halocarbon release rates, (b) the formulation of photochemical and dynamical processes in the models, (c) long term trends in other photochemically active trace gases, (d) long term trends in gases that affect the climate, and (e) trends in solar radiation.

Some uncertainties, such as deficiencies in the modelled dynamics and undiscovered processes in the atmosphere, can only be assessed by further model studies and continued comparison between model output and field measurements. Other areas of uncertainty can be assessed quantitatively, for example, the experimental uncertainties in measured photochemical reaction rate included in the models.

The effects of uncertainties in chemical reaction rates, photodissociation cross-sections and solar flux (but not in model formulation or transport parameters) have been investigated in several one-dimensional models. A Monte Carlo technique is used, whereby (typically) one thousand model runs are performed in which individual parameter values are chosen at random within an approximate probability distribution based on the assessed uncertainty of each parameter. The probability distribution for one such model with 1985 CFC emissions is shown in Figure 9.13. It is clearly asymmetrical, with a long tail towards large ozone changes. The standard deviation of the distribution is comparable to, but less than, the magnitude of the change. Estimated ozone depletion from this series of runs (with 1985 CFC emissions but all other gases held constant) was −6.2 ± 5.5%. For a series of runs with 14 ppbv chlorine, 2 times methane and 1.2 times nitrous oxide it was −7.7 ± 5.8%.

SUMMARY

Results from modelling studies with CFC, CH_4, N_2O and CO_2 emissions have provided some insight into the possible changes that may occur to the ozone layer in the future.

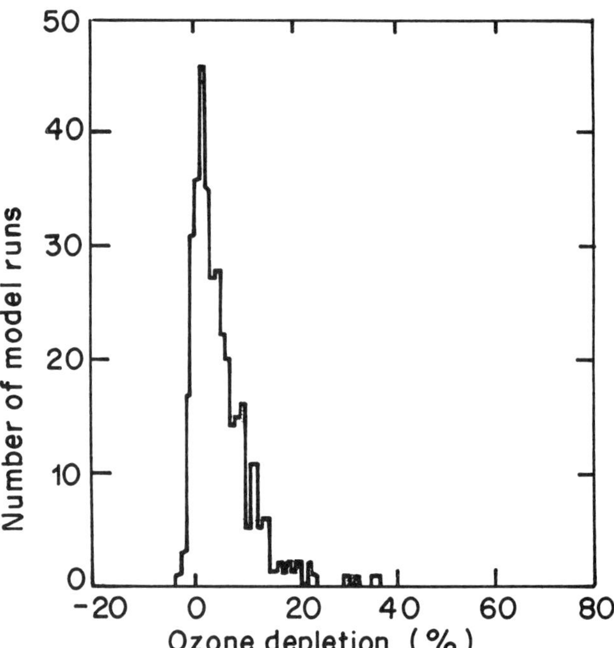

Figure 9.13 The probability distribution of percentage column ozone change predicted by a series of model runs with 1985 CFC emissions, taking account of the uncertainties in model parameters

The generally good agreement with measurements in the atmosphere might be thought to give us reasonable confidence in model predictions. However this confidence is tempered by the fact that we cannot adequately model present day ozone concentrations at 40 km, a region where the processes are thought to be relatively simple. It is therefore in the qualitative rather than the quantitative features of the calculation that greatest confidence can be placed. For instance, we have more confidence in the predictions of latitudinal and seasonal variations than in the absolute magnitude of the depletions themselves. However, even if the models were perfect, the accuracy of the predicted ozone changes would rely on assumptions made about the future trends in source gas emission. The scenarios described have assumed a continuation of present trends in CH_4, N_2O and CO_2, but this is not based on a full understanding of the sources and sinks and how they may change. CFC growth is obviously under our control, in principle.

In recent years several key issues have emerged from modelling studies of ozone change:—

(a) The model predictions of column ozone changes as a result of CFC increases are strongly dependent on simultaneous increases of other source gases, in particular those of methane, nitrous oxide and carbon dioxide.

(b) Some one-dimensional model studies have found strong nonlinearities in their results, which imply a more rapid decrease in column ozone occurring for a given increment of chlorine once the chlorine perturbation is large. This may have significant implications for ozone content if CFC emissions increase substantially. More recent two-dimensional model studies, however, have shown that nonlinearity will not become important until the levels of chlorine are very much higher than they are at present.

(c) Two-dimensional models predict large latitudinal gradients of reductions in column ozone as a result of increasing CFCs. Larger ozone reductions are predicted in mid-latitude and polar regions by the two-dimensional models than by either the one-dimensional or the globally-averaged two-dimensional models. The reason for this is that the self-healing mechanism evident at low altitudes in one-dimensional models and in equatorial regions in two-dimensional models is not significant in temperate and polar regions in comparison to other, dynamically controlled, mechanisms. Thus, global average one-dimensional model results cannot provide a reliable prediction for the depletion of ozone in mid-latitude and polar regions.

(d) Measurements of column ozone in the Antarctic have shown seasonally varying reductions of 5-40%. Current two-dimensional models predict a reduction of only 2-3% in the column ozone at these latitudes and a reduction of 15-25% in local ozone at 40 km. The large discrepancy between predicted and observed ozone reduction at these latitudes is a matter for concern and requires close attention in the immediate future.

ANTARCTIC OZONE 10

OBSERVATIONS

Climate of the Antarctic Stratosphere

The circulation in the winter stratosphere over Antarctica and the surrounding oceans is dominated by the polar vortex; a region of very cold air surrounded by very strong westerly winds. Air within the cold core of the vortex is practically isolated from that at lower latitudes. The vortex starts to form soon after the March equinox, the Antarctic stratosphere cooling rapidly as darkness spreads over the polar cap, and reaches its largest extent in mid-July. By mid-August, with the sun returning southwards, the boundary of the vortex has retreated to about 60°S and its area is then about 15% of that of the southern hemisphere. A belt of warm air occupies the mid-latitudes (30°S to 60°S). Within this belt, migrating areas of descending air (associated with the development of anticyclonic flow) can give rise to patches of very high values of column ozone. These patches are conspicuous features of maps from the satellite Total Ozone Monitoring Spectrometer (TOMS).

Prior to the mid-70's column ozone in this vortex region remained essentially constant throughout winter and early spring at about 300 DU, before increasing as the vortex broke down in early summer (the "final warming") to levels of around 400 DU. Recently this picture has changed. A new feature has been observed in Antarctica (Farman et al, 1985); the development of a deep minimum in column ozone in early spring (September/October). Values as low as 150 DU have been reported. These seasonal depletions, the first unequivocal changes in the climatology of ozone to be identified, were completely unexpected, and very divergent opinions have been expressed as to their cause and their significance for the global ozone layer.

The November 1986 supplement of Geophysical Research Letters (Volume 13, Number 12, referred to below as GRLNS for simplicity) contained 45 papers devoted to the topic of Antarctic Ozone; many of the observations and hypotheses published in GRLNS are considered in this chapter.

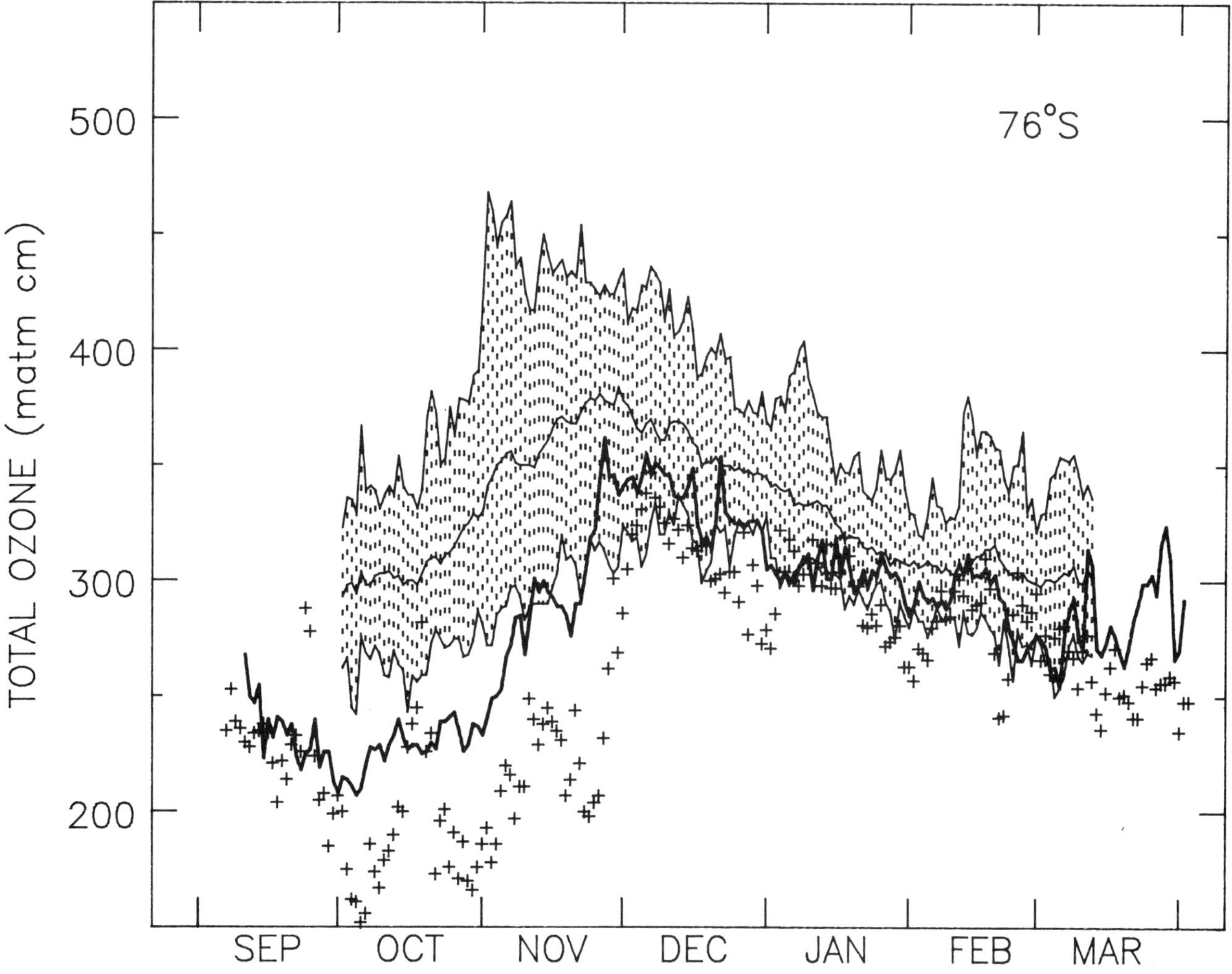

Figure 10.1 Mean and extreme daily values of column ozone over Halley Bay, 1957–73 (hatched) with mean daily values for 1980–84 (solid line) and daily values for 1985 (crosses). Units are milli-atmosphere cm. (1 matm cm = 1 Dobson unit)

Changes in the climatology of column ozone

The British Antarctic Survey has operated Dobson spectrophotometers at Halley Bay (75°S, 64°W) since 1957. Figure 10.1 shows the mean and extreme daily values of column ozone amount in 1957-1972, from September to March (observations are not possible during the polar night). The inter-annual variation is clearly greatest in November, reflecting mainly the variable timing of the final warming. It may be inferred from the extreme curves that late ozone maxima are much weaker than early ones, and this can be confirmed by inspection of the individual years.

Also shown are the mean daily values for 1980-1984, and the daily values in 1985. These show that the depletion is fastest in September. The values are consistent with there being negligible loss of ozone during the polar night, the depletion starting when the sun returns to the polar cap, for example, in mid-August at 75°S. The column ozone amount reaches its minimum values by mid-October, and shows little sign of recovery until the end of the month. Ozone amounts in the summer are lower than the mean values for 1957-73, but are not abnormally low.

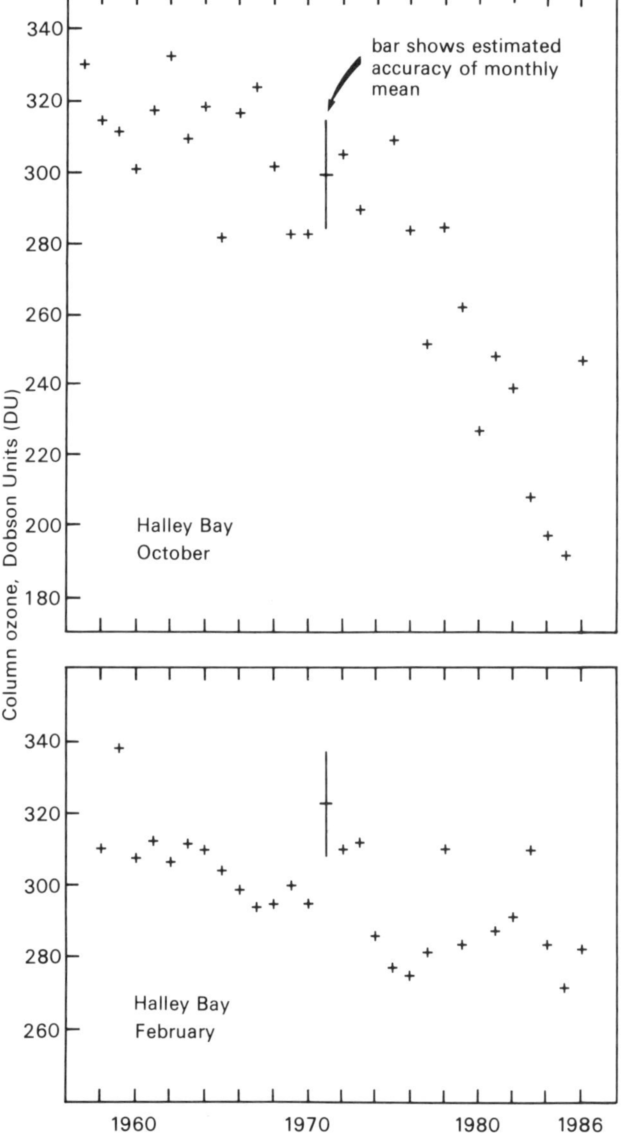

Figure 10.2 Monthly mean values of column ozone over Halley Bay for October and February, from 1957 to 1986

The monthly mean values for October and February at Halley Bay since 1957 are shown in Figure 10.2. The first significant depletion occurred in September/October 1977. The rapid fall in October mean values from 1981 to 1984 is particularly striking. There is no marked systematic change in the February values for these years. In 1985 and 1986 anomalously large fluctuations in column ozone amount were observed at Halley Bay in mid-October (Farman, 1987), raising the monthly means. However, the minimum values observed in early October were 169 DU in 1984, 152 in 1985, and 162 in 1986 (but still falling steeply when the vortex moved suddenly away from the station), so that it would be premature to conclude that the depletion had reached a limit. It should be noted that Figure 10.2 extends over practically three solar cycles, and covers two large volcanic eruptions and several major El Nino events. No significant correlation between these phenomena and the ozone record has been established.

Data from Dobson spectrophotometers at Syowa (69°S, 40°E) and South Pole stations have shown broadly similar effects (GRLNS: Chubachi and Kajiwara; Komhyr et al). The ground-based observations have been confirmed, and the spatial and temporal extents of the depletions revealed in detail, by data from the Total Ozone Mapping Spectrometer (TOMS) and the Solar Backscattering Ultraviolet (SBUV) instruments on the polar-orbiting Nimbus 7 satellite (Stolarski et al, 1986). Figure 10.3 shows TOMS maps of mean column ozone for the eight Octobers 1979-1986. The deepening of the central minimum is clearly seen. It is also apparent that there have been changes in column ozone in the warm belt with notably low mean values there in 1985.

Statistics derived from TOMS data are discussed in several papers in GRLNS (Schoeberl and Krueger; Newman and Schoeberl; Stolarski and Schoeberl; Schoeberl, Krueger and Newman), with particular attention being given to column ozone amounts in October. These papers attempt to relate these statistics to possible causes of the depletions. However, as stressed above, the most rapid depletion occurs in September.

Figures 10.4 to 10.7 show a selection of daily TOMS maps from the austral spring of 1986. It is apparent from the ozone amounts on 22 August (Figure 10.4) that losses during the polar night were small. By 2 September, eleven days later, values less than 200 DU were seen, in a rather restricted area. A month later, the region of very low ozone covered most of Antarctica. The lowest value for the year, 163 DU, was seen on 10 October, Figure 10.7. The minimum value is similar to that seen in 1984, and 20 DU greater than the minimum (143 DU) seen in 1985. However, the amounts of ozone in the warm belt in 1986

Figure 10.3 Mean monthly column ozone for the eight Octobers 1979 to 1986, from the satellite Total Ozone Monitoring Spectrometer (TOMS). (Figures 10.3 to 10.7 were supplied by NASA, courtesy Dr M Schoeberl)

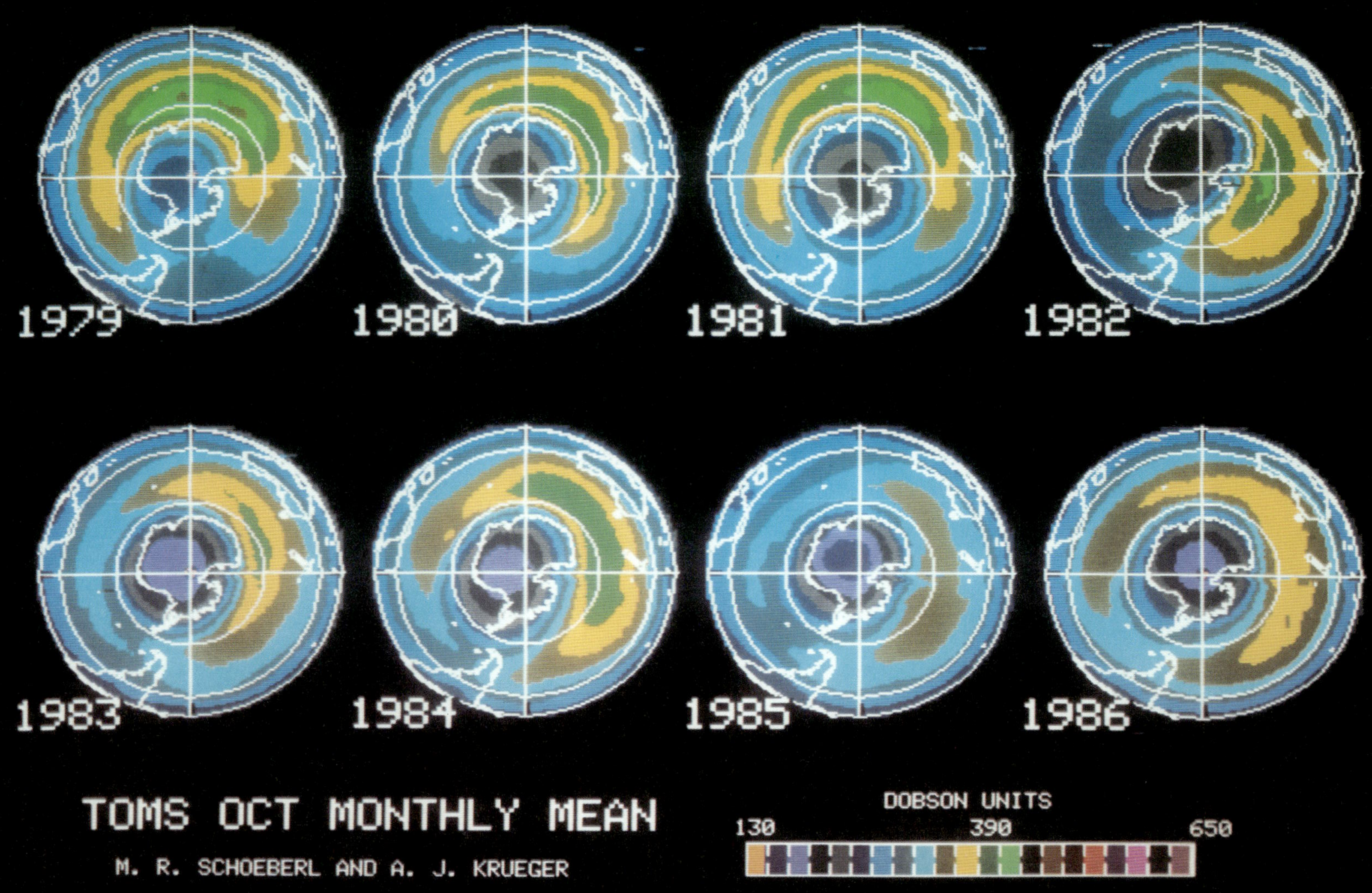

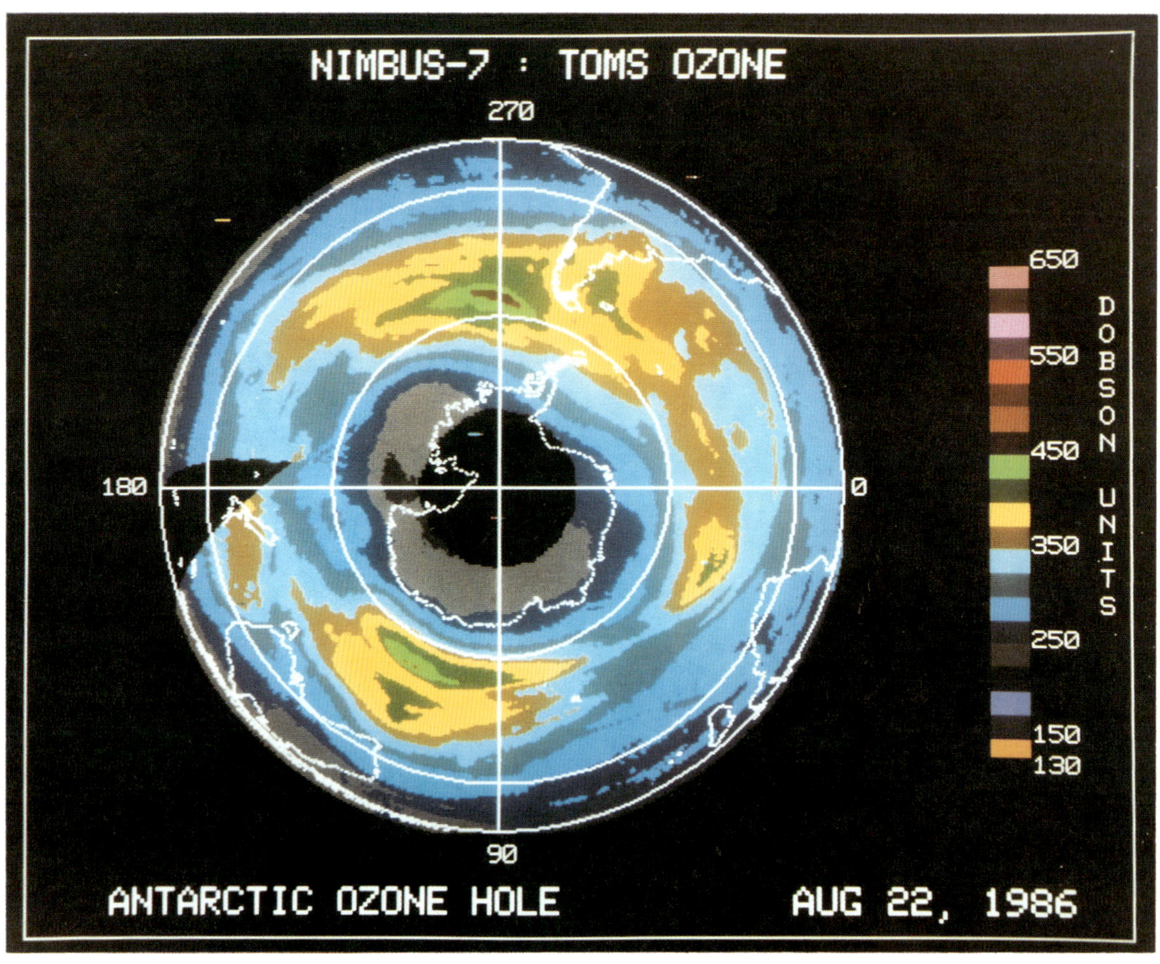

Figure 10.4 Daily mean column ozone over Antarctica for the austral spring of 1986, from TOMS. 22 August

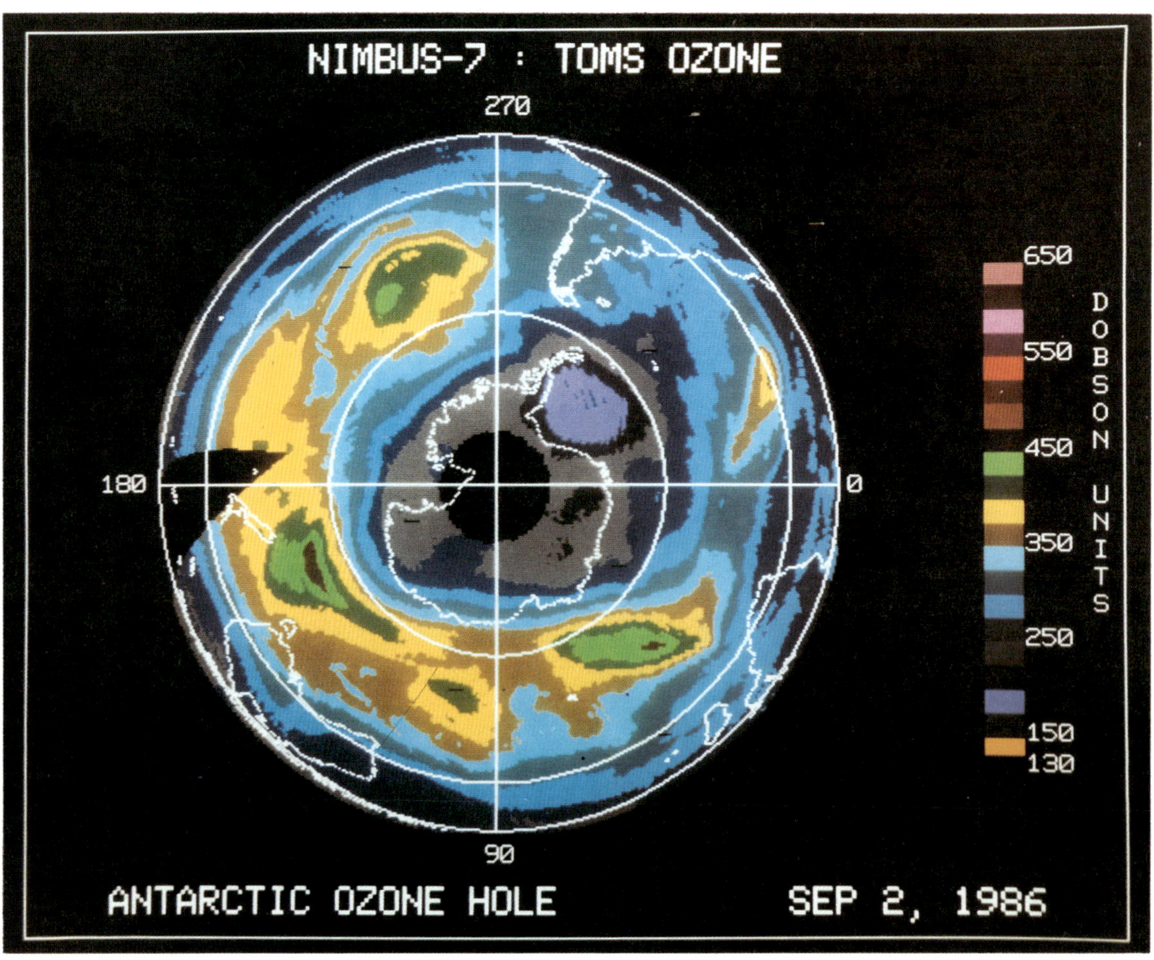

Figure 10.5 As Figure 10.4: 2 September 1986

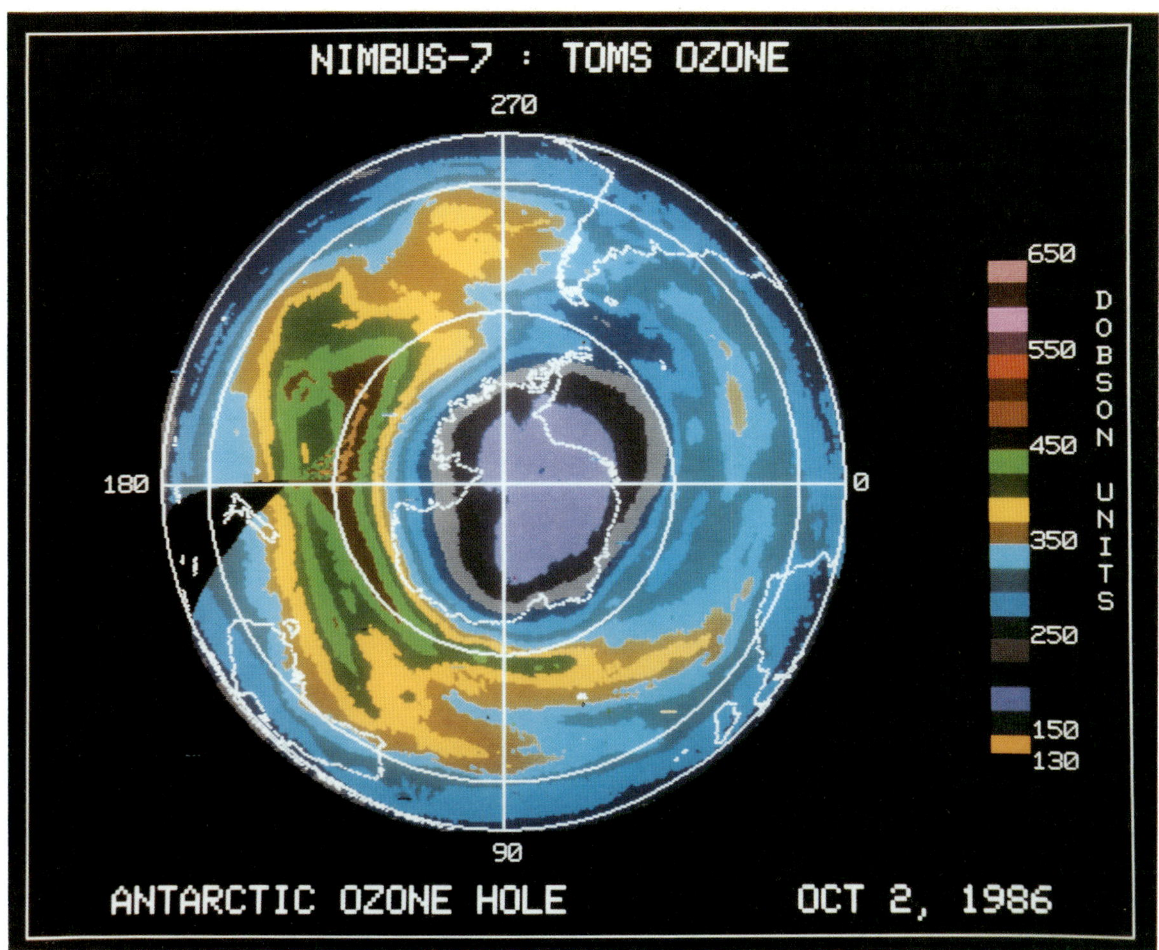

Figure 10.6 As Figure 10.4: 2 October 1986

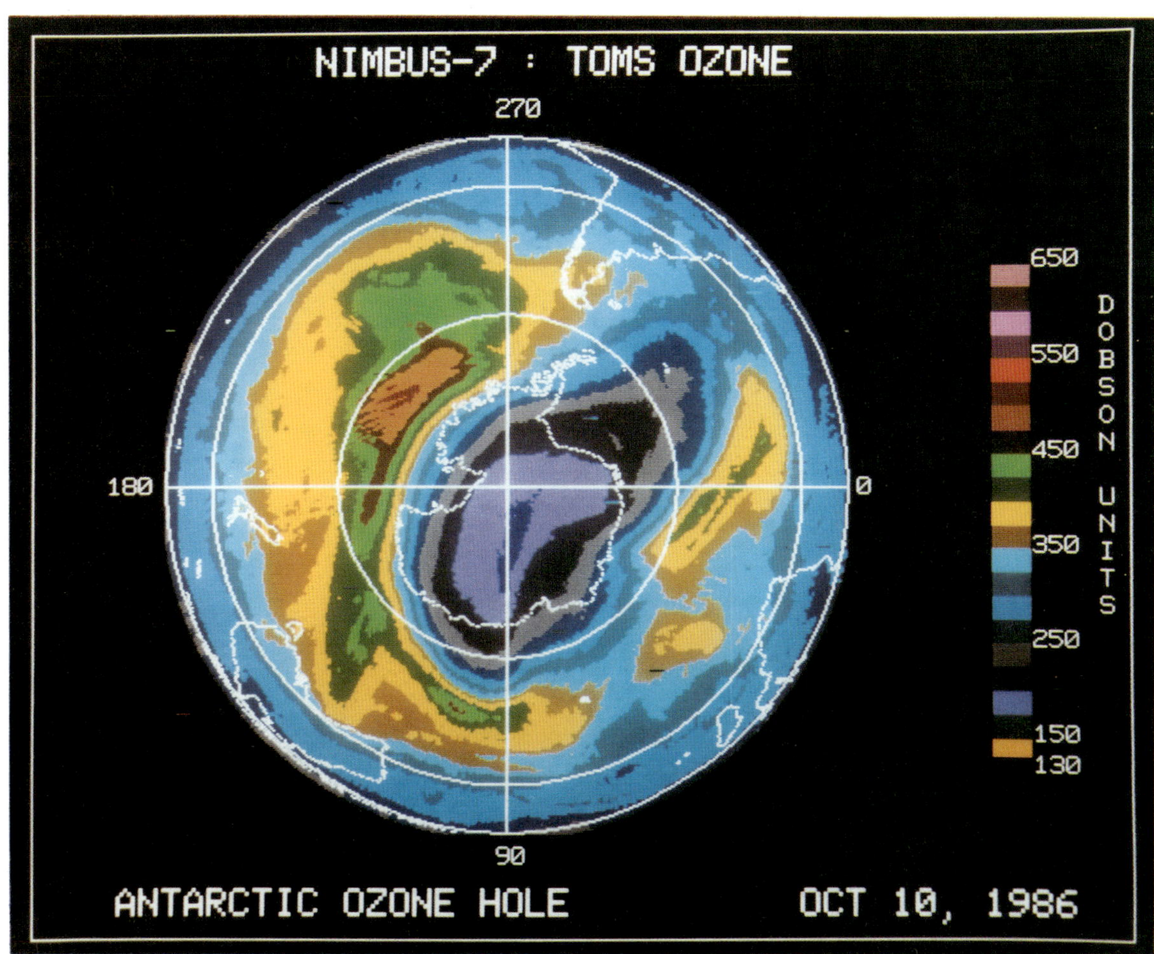

Figure 10.7 As Figure 10.4: 10 October 1986

were also considerably greater than in 1985, and it would be premature to infer that the depletions are on the wane. The ozone amount present when the vortex is formed varies from year to year, and affects the minimum value observed later. The global average of column ozone, as shown in Figure 6.2, was anomalously low in April and May 1985.

Changes in ozone profiles

Figure 10.8 shows ozone profiles, obtained by the United States National Ozone Expedition at McMurdo station in 1986 and published recently by Hofmann et al (1987). The values of column ozone obtained by integrating these profiles are 271 DU on 28th August and 155 DU on 16th October, with uncertainties of about 5%. These values are reported to be consistent with TOMS maps, which also confirm that the core of the vortex was over McMurdo on these dates. The greatest depletion has evidently occurred between about 12 km and 24 km. The lowest ozone partial pressure in the stratosphere, on 16th October, of only 10 nb, at 20 km, represents a depletion of over 90%.

The profile of the ozone mixing ratio for this sounding is shown in Figure 10.9. The decrease in mixing ratio, from 0.4 to 0.16 ppmv, in a layer some 2.5 km thick between 80 and 60 mb is decidedly anomalous. It seems highly unlikely that transport (a conservative process) has contributed to the formation of this feature, and chemistry seems to be implicated. The mixing ratios between the ground and 200 mb are also informative. A stratospheric origin for the air parcels (or tongues) around 300 and 500 mb is indicated by the enhanced mixing ratios at those levels. This is in accord with the view that the tropopause is an erosion surface, and that the direction of mean transport at that level is downwards.

Ozone-sondes are not very reliable at heights above 25 km, and it is not clear what ozone changes, if any, are occurring in the upper stratosphere. The SBUV instrument provides, potentially, coverage up to 48 km. However the retrieval algorithm has been shown to be deficient when the column ozone is less than 250 DU, and is currently being revised. McPeters et al (GRLNS) have discussed the problems, and have presented results (Table 10.1) from preliminary corrected profiles.

Table 10.1 Percentage difference from 1979 for October 7-14 each year for 75° to 80°S in the region of depleted ozone

Year	1-4 mb (38-48 km)	4-15 mb (29-38 km)	15-125 mb (15-29 km)
1980	+0.6	−6.2	−8.4
1981	+6.1	−14.3	−9.6
1982	−4.5	−1.8	−13.4
1983	−4.7	−20.1	−30.7
1984	−14.0	−16.9	−35.2

The higher two layers show great variability; only in the lowest layer is there a clear monotonic trend. Nevertheless, these data show some depletion in all layers from 1982.

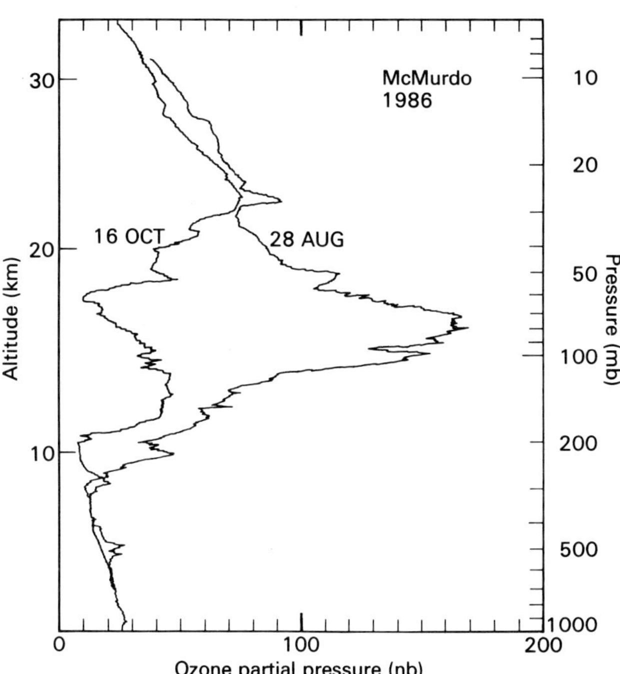

Figure 10.8 Ozone profiles from balloon soundings over McMurdo (South Pole) during 1986. (after Hofmann et al, 1987)

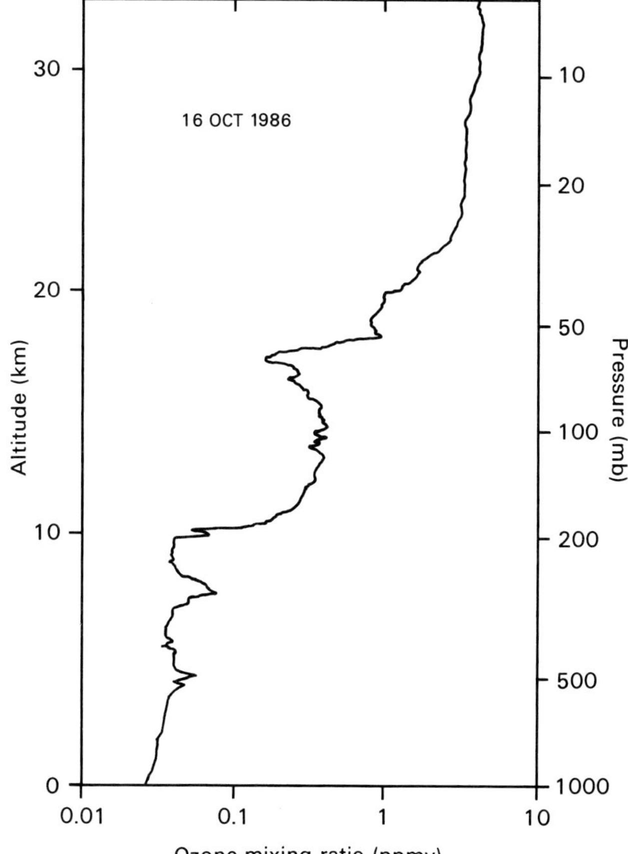

Figure 10.9 Ozone mixing ratios over McMurdo, derived from the balloon sounding of 16 October 1986 shown in Figure 10.8. (Hofmann, personal communication)

Measurements of other constituents

Ground-based measurements of NO_2 have been reported by Noxon (1978) from McMurdo, by McKenzie and Johnston (1984) and by Keys and Johnston (GRLNS). Shibazaki et al (GRLNS) report results of balloon and ground-based measurements from Syowa. There is general agreement that the vertical column density of NO_2 is very low in the spring at high latitudes.

Many early expeditions have reported the presence of clouds in the stratosphere in the Antarctic winter (for example, Liljequist, 1956). Polar Stratospheric Clouds (PSCs) are now regularly identified by satellite instruments as regions of low temperature ($<$195 K) and moderate extinction of 1 μm solar radiation. There are no direct measurements of their composition or phase or size distribution. Recent measurements have been discussed by McCormick and Trepte (GRLNS). The possible effects of these clouds on physical and chemical processes in the stratosphere will be discussed later.

Recent measurements at McMurdo

Recognising that additional measurements of trace gases were needed, the United States organised a National Ozone Expedition. Personnel and equipment were flown to McMurdo station in August 1986. Ozone and aerosol profiles were measured from balloons. The first profiles to be published have been discussed above. Measurements were made from the ground of ClO and N_2O using a microwave instrument; of NO_2, NO_3 and OClO column abundancies using an ultra-violet photometer, and of many other trace gases using an infra-red spectrometer (similar to ATMOS – see Chapter 7). The data for these trace gases have not yet been published, but some preliminary findings were announced in a Press Release (NSF, 1986). These preliminary results will be referred to in the following two sections, in the context of various ideas which have been put forward to explain the depletions.

CHEMICAL THEORIES

In their original paper, Farman et al (1985) linked the ozone reductions to increasing levels of chlorine in the Antarctic stratosphere, and this proposition has subsequently been explored by several groups using model calculations. In order to describe the observed depletion using current model chemical schemes, which predict that chlorine is tied up almost entirely as the reservoir species HCl and $ClONO_2$, the daytime concentration of the ClO radical in the lower stratosphere in the Antarctic spring needs to be substantially higher than calculated.

Several schemes have been postulated for enhancing the concentration of ClO_X radicals. Solomon et al (1986) have suggested that the heterogeneous reaction between HCl and chlorine nitrate, which has a negligible rate in the gas phase, occurs rapidly on the surface of Polar Stratospheric Clouds:

$$HCl + ClONO_2 \xrightarrow{\text{surface}} Cl_2 + HNO_3 \qquad (43)$$

This converts the NO_X component in the reservoir ($ClONO_2$) to HNO_3 and the ClO_X component to Cl_2, which is photolysed by visible light to release chlorine atoms.

The surface reaction of chlorine nitrate with water may also provide a photochemically unstable Cl source, ie hypochlorous acid (HOCl):

$$H_2O + ClONO_2 \xrightarrow{\text{surface}} HOCl + HNO_3 \qquad (44)$$

The following catalytic cycle for ozone depletion can then occur:

$$HOCl + hv \rightarrow OH + Cl \qquad (45)$$
$$Cl + O_3 \rightarrow ClO + O_2 \qquad (9)$$
$$OH + O_3 \rightarrow HO_2 + O_2 \qquad (20)$$
$$ClO + HO_2 \rightarrow HOCl + O_2 \qquad (46)$$

$$\text{net} \quad 2\,O_3 \rightarrow 3\,O_2$$

A further requirement is that NO_X concentrations are low, so that the reactions involving NO and NO_2 which counterbalance ozone removal ($ClO + NO \rightarrow NO_2 + Cl$ and $ClO + NO_2 + M \rightarrow ClONO_2 + M$) are reduced in importance. Low NO_2 concentrations continue to be an observed feature of the polar night-time stratosphere, and measurements of the vertical column of NO_2 in the "ozone hole" are the lowest yet seen anywhere (NSF, 1986). It has been proposed that these low NO_2 concentrations are due to conversion of N_2O_5 to HNO_3 via a heterogeneous reaction with water or ice:

Table 10.1 Percentage difference from 1979 for October 7-14 each year for 75° to 80°S in the region of depelted ozone

Year	1–4 mb (38-48 km)	4–15 mb (29-38 km)	15–125 mb (15-29 km)
1980	+0.6	−6.2	−8.4
1981	+6.1	−14.3	−9.6
1982	−4.5	−1.8	−13.4
1983	−4.7	−20.1	−30.7
1984	−14.0	−16.9	−35.2

$$N_2O_5 + H_2O \xrightarrow{\text{surface}} 2HNO_3 \qquad (47)$$

Acceleration of this reaction in the presence of Polar Stratospheric Clouds has been suggested. It has also been pointed out (Farman, 1987) that the additional HO and HO_2 produced as a result of the $H_2O + ClONO_2$ reaction will lead to increased ozone depletion through their direct reactions with ozone ($OH + O_3 \rightarrow H_2O + O_2$ and $HO_2 + O_3 \rightarrow OH + 2O_2$).

McElroy et al (1986) have suggested that loss of ozone in Antarctica may be attributed to a scheme in which the reaction of BrO with ClO is the rate limiting step:

$$BrO + ClO \rightarrow Br + Cl + O_2 \qquad (48)$$
$$Br + O_3 \rightarrow BrO + O_2 \qquad (11)$$
$$Cl + O_3 \rightarrow ClO + O_2 \qquad (9)$$

$$\text{net} \quad 2\,O_3 \rightarrow 3\,O_2$$

Tung et al (1986) have pointed out that an additional pathway for the ClO + BrO reaction is the formation of OClO:

$$BrO + ClO \rightarrow OClO + Br \qquad (49)$$

which would then become the dominant form of active chlorine in the polar night. The rate constant for this reaction at room temperatures is rather uncertain, and there are no data at all at stratospheric temperatures. In daylight OClO is rapidly photolysed by visible light, providing an immediate source of ClO:

$$OClO + hv \rightarrow O + ClO \qquad (50)$$

This sequence gives no net ozone depletion from the BrO + ClO reaction occurring through this channel. This reaction is the only significant source of OClO in current atmospheric models. Preliminary reports of spectroscopic measurements in the US 1986 National Ozone Expedition indicate that the OClO molecule is indeed present within the polar vortex.

Molina and Molina (1986) have suggested involvement of an additional reservoir species, Cl_2O_2, formed in the self reaction of ClO:

$$ClO + ClO + M \rightarrow Cl_2O_2 + M \qquad (51)$$
$$Cl_2O_2 + hv \rightarrow Cl + ClOO \qquad (52)$$
$$ClOO + M \rightarrow Cl + O_2 + M \qquad (53)$$
$$2\,(Cl + O_3 \rightarrow ClO + O_2) \qquad (9)$$

$$\text{net} \quad 2\,O_3 \rightarrow 3\,O_2$$

Calculations carried out by Rodriguez et al (GRLNS) using rate coefficients for the ClO + ClO reaction measured by Hayman et al (GRLNS), and the Cl_2O_2 photolysis constants estimated by Molina and Molina, show that removal of O_3 by the catalytic cycle involving Cl_2O_2 could be as important as that involving BrO.

Similar conclusions are reached by Isaksen and Stordal (GRLNS), who have used a 2D global model to estimate Antarctic depletion. They make the reasonable assumptions that stratospheric Cl has risen from 1.0 to 2.7 ppb and that bromine has risen from 15 ppb to 30 ppb in the past decade. They obtained enhanced depletions of ozone in the spring at high latitudes. However, this does not necessarily prove that a chemical mechanism is responsible for the depletion.

All of the above theories require the heterogeneous reaction between HCl or H_2O and $ClONO_2$ to enhance ClO_X. Crutzen and Arnold (1986) have suggested an alternative mechanism to enhance ClO_X, involving release from the reservoir species of HCl as a result of enhanced OH densities. These enhanced OH densities in the polar spring are ascribed to a cosmic ray source, augmented by coupling with CH_4 oxidation when sunlight returns, and also require the removal of HNO_3 from the gas phase. This mechanism could imply some dependence on the solar cycle.

A relationship between Antarctic ozone and the solar cycle has also been suggested by Callis and Natarajan (1986), who propose that ozone destruction is being catalysed by abnormally high levels of NO_X. This theory links the currently observed ozone depletion to the 1979 solar maximum, resulting, it is claimed, in the production of large amounts of NO at the top of the atmosphere. This theory is difficult to reconcile with the observed low NO_2 concentrations.

DYNAMICAL HYPOTHESES

Introduction

Farman et al (1985), describing the reductions in ozone in the Antarctic spring, suggested that, on the basis of unchanged lower stratospheric temperatures over Halley Bay, (in particular below 20 km in August and September) no significant changes in the atmospheric circulation had occurred since the ozone reduction began in earnest and that some other explanation, for example, chemistry, must be sought.

Since then, however, several authors have identified temperature changes in the low stratosphere (ie below 25 km) over the Antarctic, and have established the existence of strong correlations between ozone amounts and other variables of dynamical or meteorological significance which together suggest that changes in the atmospheric circulation may influence October ozone amounts.

An example of such a correlation is shown in Figure 10.10 (Austin et al (1987)). This shows southern hemisphere sea surface temperature (SST) anomalies plotted against Halley Bay values of column ozone. Preliminary model calcula-

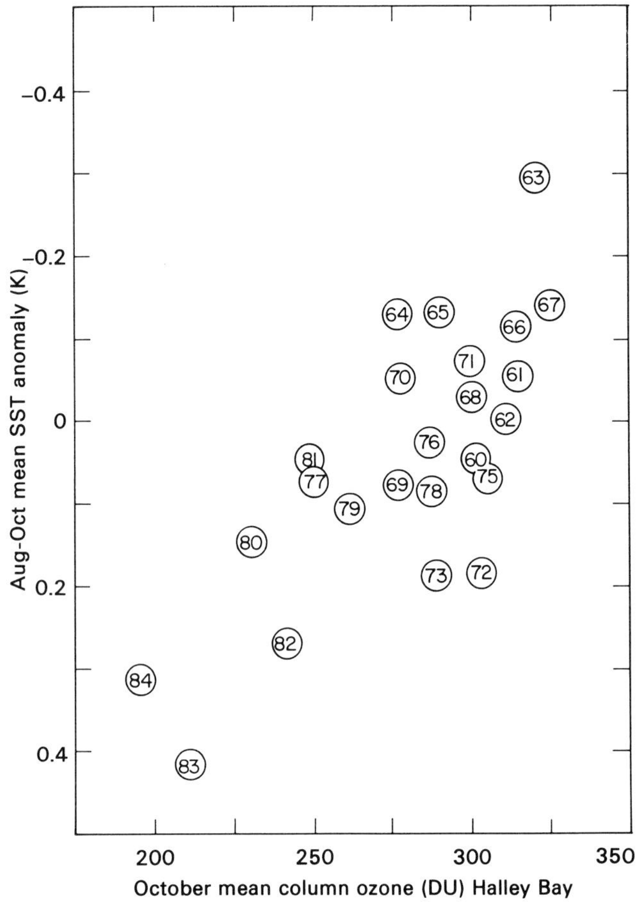

Figure 10.10 Southern hemisphere sea surface temperature (SST) anomalies polotted against values of Halley Bay column ozone

tions suggest that a negative SST temperature anomaly will reduce the strength of the equator to pole transport of ozone and thus the amounts of ozone at high latitude. A second strongly suggestive argument is that of Stolarski and Schoeberl (GRLNS) who show that the total amount of ozone between 44 and 90°S remains essentially unchanged throughout September, October and November of each year studied (Figure 10.11) suggesting that at least within each year the variations in polar ozone are determined by dynamical redistribution rather than by chemistry.

Natural variability

Detecting changes in the circulation, particularly in the Antarctic stratosphere, is hampered by the major year to year differences in the manner and timing of the breakdown of the stable winter vortex. During this breakdown low ozone concentrations (ozone-poor air) and low temperatures found within the vortex are displaced by warmer, ozone-rich air from lower latitudes.

Komhyr et al (GRLNS) and others have shown that delayed vortex breakdown (and thus delayed onset of high ozone values and warmer air) is associated with low October ozone values. However, despite the occurrence from 1980 onwards of six consecutive "late" vortex breakdowns, the timing of the breakdowns alone does not appear to be able to explain the ozone decreases seen in the Antarctic spring in recent years.

Reported changes in the Antarctic stratosphere

Newman and Schoeberl (GRLNS) report a marked cooling in the low stratosphere in October (18 K at 24 km over the south pole between 1979 and 1985). While they are cautious about ascribing a cooling of this magnitude to a secular trend (because of the large year-to-year variability) they also report a general yearly downward trend in temperature at 30 mb (24 km) averaged from 40°S to 90°S, totalling approximately 10 K between the years 1979 to 1985. They point out that the total column ozone amounts poleward of 40°S (although essentially constant during each year) have declined systematically. They argue that, since ozone and temperature changes are well correlated, the temperature changes are indicative of a change in

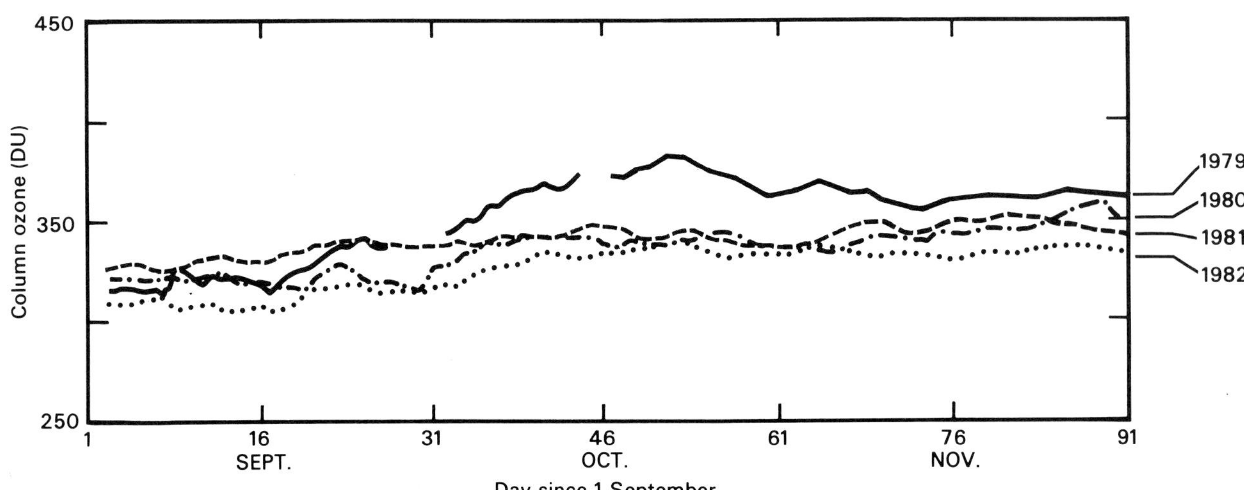

Figure 10.11 The daily area-weighted integral of column ozone from 44°S to the pole, from 1 September to 30 November, for the years 1979 to 1982

circulation (increased upward motion — see below).

Angell (GRLNS), using a different set of data, also reports (Figure 10.12) declining temperatures over Antarctica since 1979 in September-November (though not in other seasons), amounting to 6 K at 100 mb (16 km); 8 K at 50 mb (21 km) and 2K at 30 mb (24 km). Labitzke (1987) shows that the 50 mb geopotential height at the south pole is significantly lower since 1979 indicating a cooling below that level and possibly a circulation change.

A widespread, though not universal, view is that these temperature changes could not have been caused by the ozone depletion and that a change in circulation (with increased ascent within the vortex and consequential adiabatic cooling) must have occurred.

Dynamical hypotheses

To reduce ozone amounts within the vortex purely by dynamical means, the air within the vortex has to be exchanged with ozone-poor air from elsewhere in the atmosphere. Two regions of suitably ozone-poor air exist: the troposphere and the mid-latitude low stratosphere. Dynamical hypotheses proposed to date centre on the exchange of tropospheric air with air in the vortex and depend on the stratospheric air being close to radiative equilibrium.

These hypotheses, the first of which was proposed by Tung (1986), assert that enhanced heating of the lower stratospheric within the vortex in late winter, but prior to breakdown, could lead to reverse circulation with rising motion within the vortex. Such a circulation would carry ozone-poor air from the troposphere into the stratosphere and would thus reduce column ozone values. Tung suggested that the additional heating required for this to occur was due to increased amounts of aerosol following recent volcanic eruptions. Another suggestion is that the greenhouse effect has led to cooling in the low stratosphere and a consequent upwelling in the vortex.

Mahlman and Fels (GRLNS) have proposed that a change in the circulation took place in 1979 leading to a less disrupted and more zonal flow in the southern hemisphere. The latter would have the effect of producing a longer lived, colder vortex, (broadly consistent with the data of Angell (GRLNS) and Labitzke (1987)) and would have the secondary effect of producing a gradual decline in hemispheric ozone amounts as, for example, claimed by Callis and Natarajan (1986).

Mahlman and Fels argue that the colder vortex, giving an increased frequency of polar stratospheric clouds, would persist longer into the spring. Differential latitudinal heating as the sun rises will lead to regions of enhanced upward motion, drawing tropospheric air upwards and reducing column ozone. Calculations of the radiation balance in the vortex by Rosenfield and Schoeberl (GRLNS), also indicate that upwelling occurs.

Discussion

There are several aspects of the above hypotheses open to question. Pyle (GRLNS) used a two-dimensional (latitude, height) model to test the idea that local heating could force an upwelling as proposed by Tung, and found that his modelled atmospheric response was too slow to produce the observed ozone depletions. Also the assertion of Mahlman and Fels that the atmospheric circulation under-

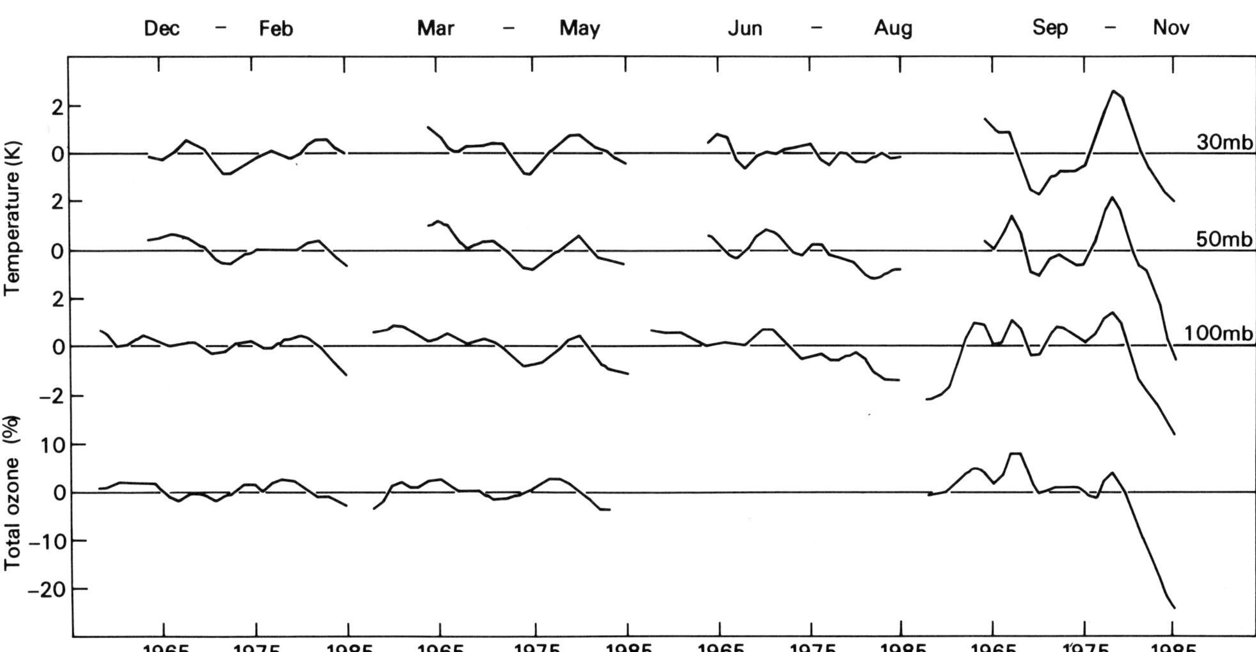

Figure 10.12 Smoothed seasonal temperature deviations at 100 mb, 50 mb and 30 mb over Antarctica, compared with smoothed seasonal column ozone deviations

went a change in 1979 is not yet established.

A number of authors have made the suggestion that the observed cooling in the Antarctic low stratosphere implies a circulation change. However, Shine (GRLNS) shows that in the absence of a circulation change, coolings of 5 K may occur as ozone is removed, a figure not inconsistent with those observed. If these radiative calculations are correct, it would imply no change in circulation.

An area which has received little attention as yet is the indirect effect of circulation changes. A stratospheric cooling could increase the frequency of Polar Stratospheric Clouds. While direct observation of these clouds by McCormick and Trepte (GRLNS) show apparently little systematic change in the extent of clouds at the relevant times of the year, any such change would affect the rates of any heterogeneous reactions.

Finally, preliminary analysis of the measurements made at McMurdo Sound (76°S) in September-October 1986 suggests that low values of N_2O were observed in the vortex. The significance of this is that N_2O is produced at the surface, is uniformly mixed in the troposphere and is destroyed in the stratosphere. If confirmed, these results imply that little tropospheric air is entering the stratosphere, contrary to the dynamical hypotheses proposed to date.

FUTURE MEASUREMENTS FROM ANTARCTICA

Plans are well advanced for a second, much larger, US measurement campaign in Antarctica, in 1987. In addition to repeating the type of ground-based observations made in 1986, aircraft will be extensively employed to make in situ stratospheric measurements for the first time. An ER-2 aircraft (developed from the well-known U-2), capable of flying through the ozone "hole" at altitudes of 20 km, will be equipped with an array of in situ sensors and detectors. A DC-8 aircraft equipped mainly with upward looking remote sounding instruments will be flown at lower altitudes. Both aircraft will be based at Punta Arenas in Chile, during late August and September 1987.

1987 will also see a considerable intensification of activity in the UK, aimed at understanding the phenomenon. The British Antarctic Survey will be making routine ozone sonde balloon flights from Halley Bay, so that the gradual depletion of the ozone can be followed with good time resolution. The Meteorological Office is collaborating in the forthcoming US measurement campaign. One component of this will be forecasting trajectories of air parcels in the vortex to allow the US aircraft to intercept the same air mass on several occasions so that chemical changes can be detected. Numerical chemical models developed at the Meteorological Office will also be used to test chemical hypotheses of the ozone depletion.

At Cambridge University, the existing two-dimensional stratospheric chemistry model will be modified to study Antarctic chemistry in detail. At Harwell, the rates of some of the novel reactions which have been proposed will be measured at temperatures close to those of the polar stratosphere.

It seems likely therefore, that knowledge of the Antarctic phenomenon will improve substantially in the next one or two years. Although major uncertainties will still exist, this should enable a better assessment to be made of the implications for ozone over the rest of the globe.

POSSIBLE OZONE DEPLETION OVER THE NORTHERN HEMISPHERE

Reports have recently appeared (New Scientist, 1986; Chemical and Engineering News, 1986) outlining the results from an analysis, carried out by Heath, of the Nimbus 7 SBUV ozone data. These results have not appeared in the reviewed scientific literature, and thus far have not been substantiated. Heath claims that, between the years of 1978 and 1984, there has been a decrease of about 3% in globally averaged, annual mean, column ozone. The decrease has not been uniform over the globe, however, with the areas of largest column ozone depletion being over the Arctic and the Antarctic.

It is difficult to comment on these findings until the full details are available in the open literature. Small trends such as these are difficult to identify from satellites. Furthermore the accuracy of the satellite data may have been affected by the increased level of stratospheric aerosol following the El Chichon volcano of 1982.

NASA have recently established a panel of scientists and asked them to scrutinise the large quantity of ozone data, ground-based and satellite, which has been gathered over past years. It is hoped that they will be able to establish what the long term trends are in ozone, with more certainty than at present. The scientific panel is expected to report its findings at the end of 1987.

CHEMICAL COMPOSITION, RADIATION AND CLIMATE

11

THE GREENHOUSE EFFECT

The earth-atmosphere system is heated by solar (shortwave, visible) radiation; this heating must be balanced by cooling due to terrestrial (longwave, thermal) radiation emitted to space. The amount of terrestrial radiation emitted is proportional to the fourth power of the temperature of the earth. If the earth had no atmosphere, the surface temperature that it would have to have in order to balance the amount of solar radiation absorbed would be 225 K (−18°C).

Certain gases (water vapour, carbon dioxide, ozone, methane, CFCs, nitrous oxide) absorb solar radiation weakly, but strongly absorb and emit longwave radiation. The (increased) downward longwave radiation from these gases is responsible for the earth's surface temperature being 288 K (15°C), some 33 K warmer than that with no atmosphere, a phenomenon often referred to as the 'greenhouse'' effect. The transfer of shortwave and longwave radiation within the earth-atmosphere system is shown schematically in Figure 11.1.

Most of the radiation emitted to space emanates from the atmospheric gases rather than the surface. The effective emitting temperature of 255 K corresponds to a height of about 6 kilometres. By increasing the concentration of an atmospheric absorber such as CO_2, the mean level from which radiation escapes to space moves to a higher, and therefore colder, level.. The longwave cooling to space is reduced and the earth/atmosphere system warms until the longwave cooling again balances the incoming solar radiation.

It has been recognised for some time that the gradual increase in carbon dioxide concentrations observed in past decades in the atmosphere could, through the greenhouse effect, lead to increases in surface and tropospheric temperatures. More recently, however, it has been realised that other absorbers, many of them man-made, although present in the atmosphere in much smaller concentrations, can produce temperature changes comparable with those due to carbon dioxide.

The addition of trace gases to the atmosphere may have other effects than the direct thermal influence just described. Thus, as a result of chemical reactions, trace gases can modify the concentrations of other, radiatively active gases. For example, CFCs may affect the concentration of ozone. Additionally, there are important feedback mechanisms which enhance the direct heating effects. Thus, a warmer atmosphere generally contains more water vapour, which is itself a greenhouse gas. Other feedback mechanisms are due to changes in sea-ice amounts, the atmospheric lapse rate (the rate of change of temperature with height) and average cloud cover.

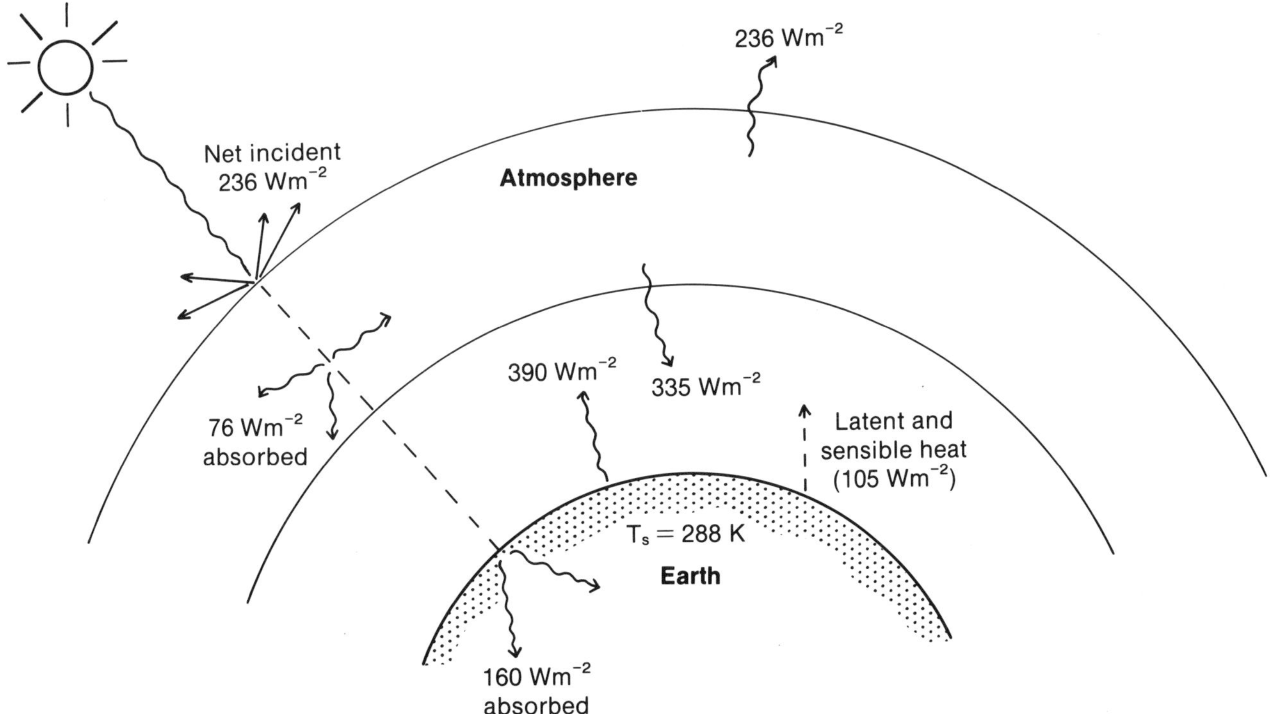

Figure 11.1 Fluxes of solar (shortwave) and terrestrial (longwave) radiation within the earth-atmosphere system. The earth's surface absorbs 160 Wm^{-2} solar radiation, plus 335 Wm^{-2} infra-red radiation from gases and clouds. To balance this input, the surface radiates 390 Wm^{-2} (appropriate to a surface temperature of 288 K); the transfer of heat and water vapour from the surface to the atmosphere accounts for the remainder

RADIATION

In order to understand the effect of an increase in a particular absorbing gas, the spectral dependence (ie the dependence on wavelength) of emission and absorption in the atmosphere must be considered.

As explained earlier, the earth receives radiation from the sun's photosphere, which is at a temperature of approximately 5800 K, and re-emits at temperatures between 180 K and 300 K. The incoming solar radiation is therefore at shorter wavelengths (in the visible at around 600 nm) than the re-emitted terrestrial radiation (which peaks in the infra-red near 13 μm) (Figure 11.2). This means that the incoming and outgoing radiation fluxes lie in different regions of the spectrum and thus are susceptible to different modifying influences. Atmospheric absorption in the region of the outgoing radiation flux is dominated by that of carbon dioxide and water vapour (Figure 11.3). Their combined effect leaves a region of high atmospheric transmission between 7 μm and 13 μm through which 70-90% of outgoing radiation escapes to space. This region is termed the atmospheric "window". Gases such as O_3, CH_4, N_2O and CFCs absorb within this "window' region and are therefore particularly effective as absorbers of outgoing radiation and thus as greenhouse gases. It has, for example, been calculated that one molecule of CF_2Cl_2 (CFC 12) is as effective as approximately 10,000 molecules of carbon dioxide, whose bands are almost saturated, demonstrating that gases present in comparatively small concentrations can still be important greenhouse gases.

Ozone differs from other gases in that it absorbs in both the ultra-violet and the infra-red, attenuating the incoming solar radiation as well as modifying the outgoing infra-red. The attenuation of solar radiation reaching the surface depends only on the total number of ozone molecules in the atmospheric path while the modification of the outgoing infra-red radiation depends critically on their vertical distribution. In the low and middle troposphere, ozone behaves as a conventional greenhouse gas, any increase leading to a surface warming. A reduction in the low stratosphere would increase the outgoing infra-red radiation, leading to a reduced surface warming or even a surface cooling. Such details of the ozone changes in the lower stratosphere and upper troposphere cannot be resolved adequately with present modelling capabilities.

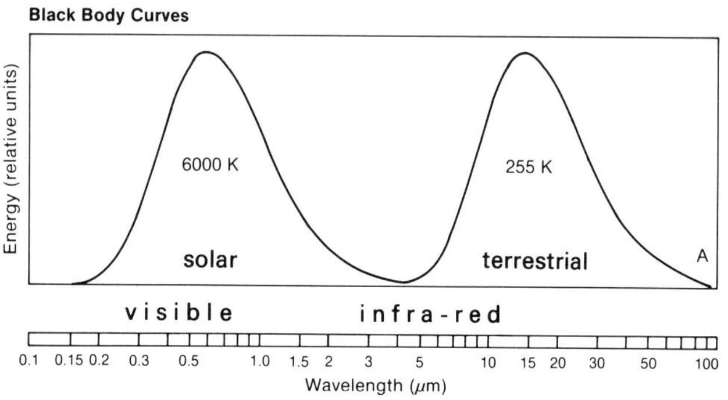

Figure 11.2 Wavelength distributions of solar and terrestrial radiation. The almost complete separation of the two wavelength ranges allows their transmission through the atmosphere to be treated independently

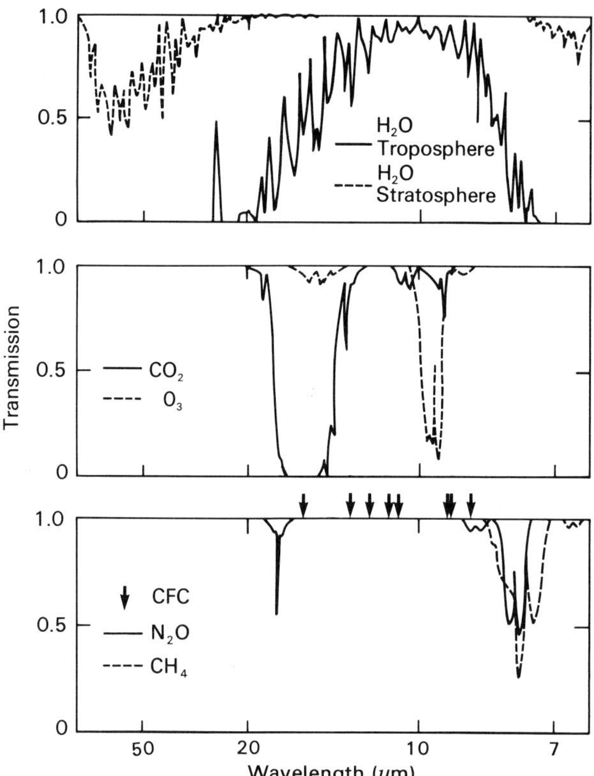

Figure 11.3 The transmission of the atmosphere in the infra-red, showing how outgoing terrestrial radiation is absorbed by the various trace gases. The arrows indicate absorption due to CFCs. (after Lacis et al, 1981)

ESTIMATING CLIMATE CHANGE

Models

The direct effects and the exploration of some feedback mechanisms can be studied using radiative-convective models (RCM). These are 1-dimensional, representing a vertical column of the atmosphere, and they calculate the atmospheric thermal structure which, for a given solar input and atmospheric composition gives radiative balance.

A vertical convective term is included to maintain an approximate correct atmospheric structure by creating a characteristic tropospheric lapse-rate in the lower part of the atmosphere. Such models are economical in computing resources and have therefore been extensively used in exploratory investigations.

For more complete simulations of the effect of greenhouse gases in climate, it is necessary to use global general circulation models (GCMs). They are three-dimensional, similar to numerical forecasting models, and, in principle, are capable

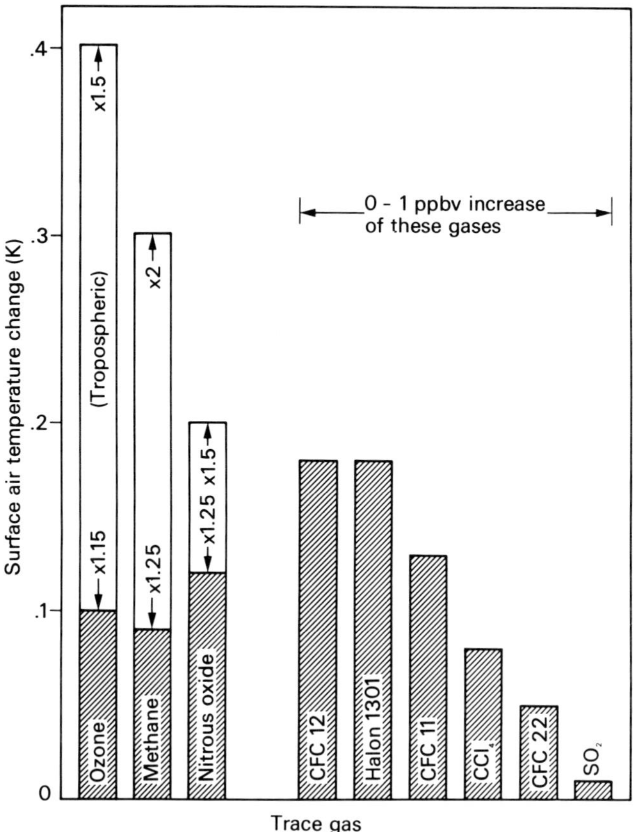

Figure 11.4 The relative effect on surface air temperature of indicated increases in ozone, methane and nitrous oxide, and of a 0-1 ppbv increase in six other trace gases

of including all the important physical mechanisms which might cause, or result from, greenhouse gas warming. In practice, however, they are subject to a number of significant limitations. Their computational demands are heavy, and the use of low resolutions or gross simplifications of the physical processes, which partially relieve this problem, limit the accuracy of the model simulations. More fundamentally, there are numerous scientific uncertainties, where the models have to proceed on the basis of present knowledge, which cannot be removed without a great deal more research. These affect particularly the main feedback mechanisms, of which the changes in ocean circulation, in sea-ice cover, in cloud amount, in vegetative cover and in other indirect biological effects can be singled out as especially significant.

Model calculations

Given suitable spectral data and a knowledge of atmospheric abundances, direct radiative effects can, at least conceptually, be evaluated for different gases. For example, Figure 11.4 shows the direct radiative effects on surface temperature predicted as a result of 0 to 1 ppbv changes in concentration of these gases.

Ramanathan (1985) has used a radiative convective model to compute the net surface temperature increase due to assumed increases in trace gas concentrations between 1980 and 2030 (see Table 11.1). The assumed rates of increase for gases such as CH_4 are based on present day trends which are not well understood. The RCM model included some feedback effects directly, and account was taken of feedback mechanisms not included by multiplying the computed temperature changes by a factor of 1.5. In Figure 11.5 is shown the surface warming predicted by this study. Vertical bars associated with each gas show the cumulative uncertainty due to variations in assumed release rates. While the cumulative surface temperature change ($\sim$ 1.5 K) is somewhat uncertain, the predicted effects of increases in gases other than CO_2 (see right hand ordinate) is seen to be comparable to that due to CO_2 alone. This study shows that trace gas increases, acting either directly via their radiative effects, and/or acting by indirect means, could significantly change temperatures at the surface and indeed at upper levels (see Figure 11.6).

As already noted, more realistic simulations of global climate require the use of three-dimensional global general circulation models. Mostly, modellers have attempted to estimate the effects of doubling or quadrupling the concentration of carbon dioxide and have not included other trace gases explicitly. However, it is reasonable in calculating the thermal effects to translate the concentrations of well-mixed greenhouse gases into equivalent concentrations of carbon dioxide, and, in this way, the results

Table 11.1 The estimated abundance of greenhouse gases, 1980 and 2030,used to calculate the relative global warming shown in Figure 11.5

Greenhouse gas	Global average mixing ratio (ppbv)			Basis of most probable estimate
	1980	2030 Most probable	2030 Range	
Carbon dioxide	339×10^3	450×10^3		2.4% per year rise in emissions
Methane	1650	2340	1850–3300	0.7% per year rise
Nitrous oxide	300	375	350–450	0.45% per year rise
CFC 11	0.18	1.1	0.5–2.0	3% per year rise in emissions
CFC 12	0.28	1.8	0.9–3.5	3% per year rise in emissions
Ozone (tropospheric)				0.23% per year rise
Ozone (stratospheric)	based on modelled changes of between +3.8% (10 km) and -38% (40 km)			

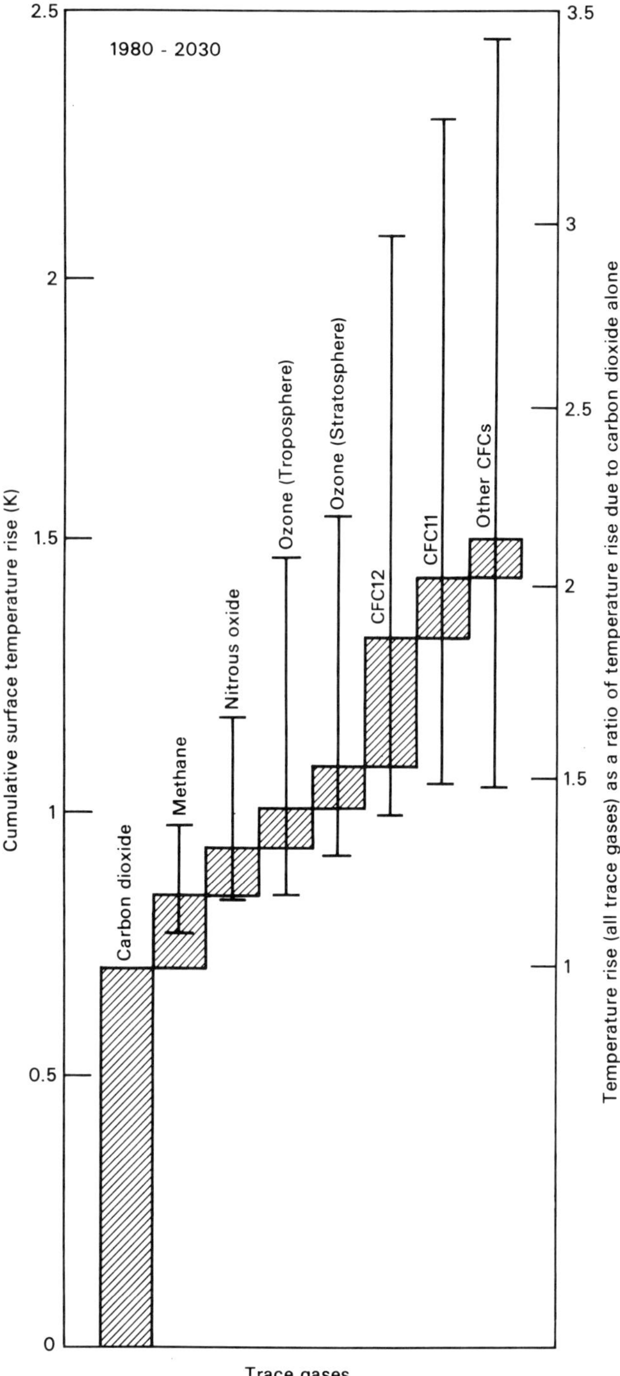

Figure 11.5 Cumulative equilibrium surface warming predicted by a radiative convective model, due to assumed increases in the trace gases (Table 11.1). The vertical bars show the cumulative uncertainty associated with release rate projections

can be interpreted as including other greenhouse gases implicitly. Note that there may be particular difficulties with ozone (which is not well mixed and is active at solar as well as infra-red wavelengths).

The first carbon dioxide assessment produced by the US National Academy of Sciences in 1979, gave an estimate of the steady state global warming due to a doubling of carbon

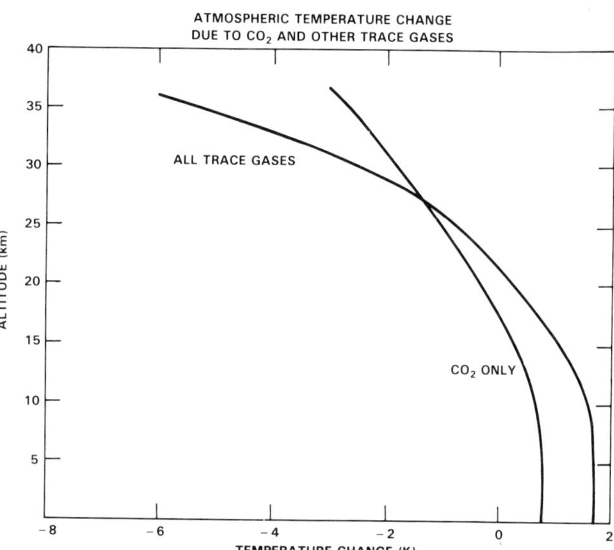

Figure 11.6 Atmospheric temperature change due to CO_2 and other trace gases, as a function of altitude

dioxide (from 300 ppmv to 600 ppmv) of 3 ± 1.5 K. The values were a judgement based on all the model results available at that time. More recent assessments (SCOPE, 1986; NASA/WMO, 1986) have seen no reason to revise the figures in the light of recent results. It is to be noted, however, that the figures are considered to define the 50% limits; that is, there is a 50% probability that the actual global warming will be greater than 4.5 K or less than 1.5 K. This very large uncertainty is, as has already been stressed, due to the difficulty of estimating the feedback mechanisms which can have a very large influence on the final warming estimate. Considerable research is being carried out by modellers into three of the most important feedback mechanisms; namely, those due to cloud, sea-ice and the oceans.

Because of the severe problems involved in simulating cloudiness realistically, early simulations assumed that cloud did not change as a result of carbon dioxide enhancement. More recently models have been able to calculate cloud internally, and therefore to allow for the feedback due to changes in cloud distribution, vertically as well as horizontally.

A number of model runs have obtained results indicating an approximate doubling of the previously estimated warming, as a consequence of making the clouds interactive. Despite the measure of agreement, this is not a result that can be accepted with confidence, particularly because all models have used a process of "tuning" or optimisation to achieve acceptable cloud parametrisations and it is not proven that the parametrisations are appropriate for an atmosphere warmed as a result of high carbon dioxide concentrations. Furthermore, it has been suggested that the optical properties of clouds would change in such a way as to reduce the warming by as much as 50%. Research on cloud parametrisations, cloud optical properties, and the verification of cloud simulations, is proceeding.

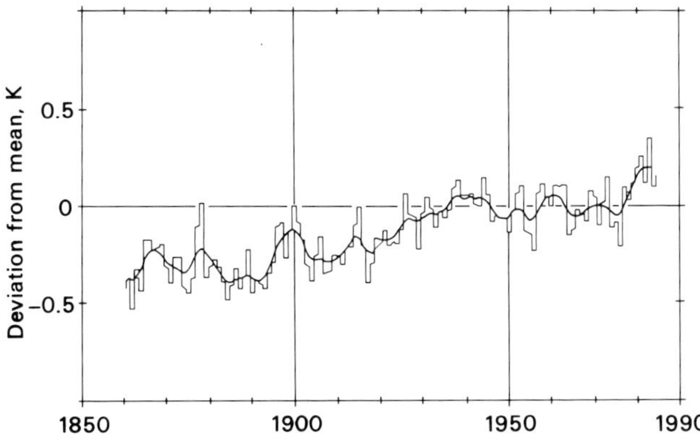

Figure 11.7a Changes in global average land surface air temperatures over the last hundred years, as analysed by Jones et al, 1986

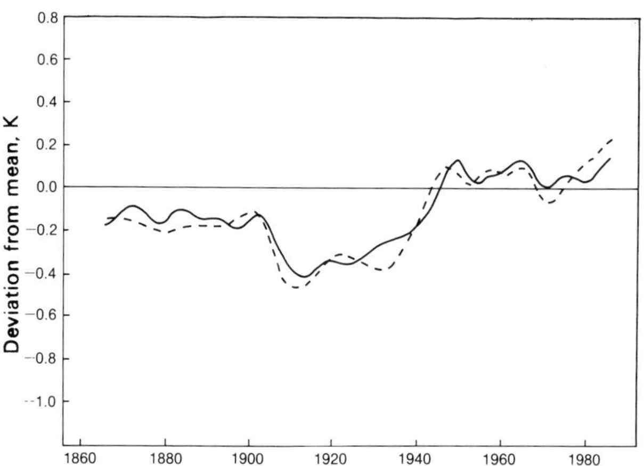

Figure 11.7b Changes in global average marine surface air temperatue over the same period as Figure 11.7, as analysed by Parker (1986). The effect of changes in global data coverage on the time series of marine air temperature is shown by the two curves; the solid line uses all data, the dashed line is restricted to 5° latitude x longitude areas with at least 90% of seasons having data in 1861-1870

The simulation of sea-ice has been achieved by sea-ice models within atmospheric models. The sea-ice models used so far are rudimentary and considerable problems have been encountered in producing realistic simulations of the present distributions near both poles simultaneously. The models are, at this stage, purely thermodynamic, taking no account of movement of wind or of internal stresses. More ambitious sea-ice models are under development.

The representation of the ocean in most climate change simulations has been highly simplified, with either no, or a very crude, allowance for the effect of changes in ocean circulation. It is obviously essential to use fully coupled ocean-atmosphere models, but the development of such models has not reached the stage where their simulations of the present global climate are acceptable.

All general circulation model simulations have so far determined the steady-state climate resulting from a change in greenhouse gases; that is, they have assumed sufficient time to reach a thermal balance and have ignored transient effects. The oceans, however, would warm only slowly, and their most important effect may be to delay the realisation of the expected warming. This delay, on present estimates, is likely to be a few decades. Thus the "realised" (actual) surface warming in 2030 is expected to be somewhat less than the equilibrium warming shown in Figure 11.5. Another consequence of the lag is that if trace gas concentrations were held at today's values, the atmosphere/surface would continue to warm as the "unrealised" temperature increases become "realised" over the next 10-100 years, ultimately reaching its equilibrium value.

OBSERVED TRENDS IN TEMPERATURE

Records exist for surface air temperature back to the mid-19th Century. Two UK groups have analysed these records, but there is an important and continuing difference between their results. The most recent analysis of land surface air temperatures by the group at the University of East Anglia (UEA) show the coolest years occurred prior to 1900 and the warmest since 1940 (Figure 11.7a; Jones , 1986). In particular they comment that the warmest three years in the record have occurred in the 1980s. On the other hand analyses of marine surface air temperatures carried out by the Meteorological Office (Figure 11.7b; Parker, 1986) show temperatures in the mid- to late 19th Century that are much closer to present day values than in the UEA analysis.

The difference in the analyses may be real, but they can only be accepted if other possible causes, such as changes in instrumentation or observing practice, can be discounted. Investigations into this difference are being carried out by the two research groups. One investigation has already indicated that as far as marine temperatures are concerned, the change in data density from 1860 to the present day is probably not a relevant factor. This is illustrated in Figure 11.7b.

Resolution of the difference between these temperature analyses is important, since it has a crucial bearing on the reference period used to detect climatic change induced by man's activities. Also, since other mechanisms may also influence climate and change global temperatures (such as changes in deep sea ocean circulation or variability in solar radiation) it is important to establish the magnitudes of any natural decadal and longer timescale temperature variations which would otherwise mask trends due to changes in greenhouse gases.

UK RESEARCH PROGRAMMES 12

ATMOSPHERIC MEASUREMENTS

From the ground

Dobson spectrophotometers, which measure the overhead column of ozone, continue to be operated at Bracknell and at St Helena in the South Atlantic by the Meteorological Office, and at the Argentine Islands and Halley Bay by the British Antarctic Survey (BAS). Following the unexpected discovery of a large decline in ozone above Halley Bay in the spring, many groups from many countries have sent instruments to Antarctica. BAS installed an NO_2 spectrophotometer on loan from New Zealand's DSIR.

A new instrument to measure the overhead column of stratospheric NO has been built at Oxford University. The radiometer, which measures solar absorption by the novel technique of Zeeman modulation, operated at Pic du Midi during the 1985 MAP-GLOBUS campaign.

As part of a European collaboration to study tropospheric ozone, the University of East Anglia will measure the seasonal variations of ozone, oxidants (hydrogen peroxide and peroxyacetyl nitrate) and source gases, both at a site in the Northern Pennines and from BAS ships.

At the University of Aberystwyth, stratospheric temperatures and aerosols have been measured by lidar. In 1982, the evolution of the dust-cloud from El Chichon was monitored. Future improvements to the facility may permit measurements of stratospheric ozone.

At the National Physical Laboratory, a new radiometer is planned which would measure minor constituents in the stratosphere. By looking at the absorption of sunlight in the infra-red and by using a laser-heterodyne detector, lines of atmospheric absorbers can be measured with very high sensitivity and spectral resolution. The resolution is such that tropospheric absorption can easily be distinguished from stratospheric, and further information about the vertical distribution of the absorber in the stratosphere can probably be obtained.

From aircraft

New measurements of minor constituents have been made from the Meteorological Office's Hercules, in the NE Atlantic near jet streams. The results are being analysed, together with satellite data, to help our understanding of stratospheric-tropospheric exchange.

From balloons

The Meteorological Office has constructed instruments to attach to a manned balloon intended to fly around the world in the southern sub-tropical jet. These will measure O_3, H_2O, temperature, pressure and wind-shear and will be lowered by winch down to 1 km below the balloon.

The new version of Oxford University's Balloon-borne Pressure Modulator Radiometer, which measures profiles of NO, NO_2 and N_2O_5, had its first flight during the 1985 MAP-GLOBUS campaign. Results are still being analysed. Results from all flights of the old radiometer (destroyed in 1983) have been compared to predictions by RAL's photochemical model. The results suggest that NO_2 falls during the night, consistent with predictions of N_2O_5 production. Improvements are now being made to the new radiometer which will greatly improve its performance.

In a collaborative programme with the Max-Planck Institute at Lindau and with KFA Julich, stratospheric air collected by MPI's balloon-borne cryogenic sampler has been analysed at Harwell. Many new hydrocarbons and halocarbons were detected and measured, including bromine source gases.

From satellites

At Oxford University, measurements of temperature, CH_4 and N_2O from SAMS on Nimbus 7 have been compiled into reference atmospheres for COSPAR. The unexpected double-peaked structure of CH_4 and N_2O has now been satisfactorily reproduced in a model (see Chapter 8). At Oxford and RAL, SAMS measurements of CH_4 were combined with LIMS measurements of H_2O to show that the total ($H_2O + 2 \times CH_4$) is constant over a wide range of latitude, height and season.

In April 1985, the Jet Propulsion Laboratory's ATMOS interferometer flew on Spacelab 3. From 19 solar occultations, it produced the first high-resolution spectra of stratospheric constituents from space. This wealth of data is being analysed by many groups around the world. Oxford University and RAL have produced preliminary profiles of rotational temperature, CH_4, H_2O, N_2O, NO and O_3; and upper limits of HOCl and H_2O_2.

The Meteorological Office continue to analyse stratospheric temperatures from SSU's on the NOAA series of operational satellites. A study of the seasonal evolution of temperature and potential vorticity is nearly complete. It emphasises the southern hemisphere, where there are few non-satellite measurements.

The Improved Stratospheric and Mesospheric Sounder (ISAMS) is now being built at Oxford University, RAL and British Aerospace. It will measure temperature, H_2O, CH_4, CO, N_2O, NO, NO_2, N_2O_5, HNO_3 and O_3. It will be launched on NASA's Upper Atmospheric Research Satellite (UARS) in 1991.

RAL and Herriot-Watt University are collaborating to build the H_2O channel to be incorporated into the Jet Propulsion Laboratory's Microwave Limb Sounder. The MLS will also measure ozone, ClO and H_2O_2 and will also be mounted on UARS.

LABORATORY MEASUREMENTS

Spectroscopy

RAL's BOMEM interferometer has provided spectroscopic data for many stratospheric constituents: HCl, HF, H_2O and CH_4 in support of HALOE on the Upper Atmospheric Research Satellite; CH_4 and NO in support of ISAMS, also on UARS; and HNO_3 and $ClONO_2$ in support of ATMOS. The interferometer measures to very high resolution (0.002 cm^{-1}) and uses a wide range of absorption cells (1 mm to 500 m), some of which can be cooled to 200 K. Spectral line strengths and widths are derived by computer, and for most gases the self-broadened widths were measured as well as the temperature-dependence of nitrogen-broadened widths. In collaboration with Oxford University's Department of Physical Chemistry, spectra of N_2O_5, $ClONO_2$, NO_3 and ClO are now being measured.

Chemical kinetics

At Cambridge University, by combining flash photolysis with infra-red diode-laser spectroscopy, the reactions and spectrum of HO_2 have been studied at the pressures of the lower stratosphere. The rate of recombination of HO_2 was published, and the calculated strength of its v3 band was shown to be a factor of 7 too high. In the future, the band strengths of NO_3 will be measured; as will the pressure dependence of the reactions HO_2 + NO and HO_2 + NO_2, by means of the difficult but sensitive technique of laser-magnetic-resonance spectroscopy in the mid-infra-red.

At Harwell, the reactions of HO_2 with Cl and ClO have been studied at room temperature by molecular modulation. The rate coefficients did not vary with pressure, although the pressure dependence of the second reaction is still in doubt at low temperatures. Using the same technique the reaction ClO + ClO is now being investigated.

Also at Harwell, a stopped-flow kinetic spectrometer was used to investigate the reactions of $ClONO_2$ with HCl, HBr and H_2O by observing its decay in the presence of excess H-containing reagent. A photo-diode array was used to obtain wide-band absorption spectra and cross-sections for $ClONO_2$. The results showed good agreement over the whole wavelength range with the lower of the values previously reported.

At Oxford University, oxidation reactions of CFC radicals, and reactions of OH and of $O(^1D)$ have been investigated. The absolute absorption cross-section of NO_3 has also been measured by several independent techniques. In collaboration with Harwell, the reactions of NO_3 with Cl, ClO and HO_2 have been investigated.

MODELS

The Meteorological Office and Oxford University three-dimensional model

This model of the stratosphere and mesosphere has high spatial resolution. The major stratospheric warming of the northern winter 1984-85 has been reproduced closely in a 10-day forecast with the model. The dynamics of the event are being investigated, in particular the non-conservative role of radiation. The model is also being used in idealised experiments to study the effect on the stratosphere of large-scale features in the tropospheric circulation. The representations in the model of radiative processes and of gravity wave breaking are being improved. Simple ozone chemistry will be included and expanded. Predictions by the model will be compared to satellite measurements of minor constituents, particularly from UARS.

The RAL and Cambridge University two-dimensional model

The model's ability to reproduce satellite data has been improved by including a treatment of equatorial dynamics. In particular it has been shown that the treatment of the semi-annual oscillation is important in determining the distributions of long-lived gases. The model reproduces the observed double-peaked distributions of CH_4 and N_2O. Predicted latitudinal and seasonal changes in O_3 and in NO_2 at night are also consistent with the observations.

Changes in O_3 and temperature, due to increasing CFCs, CH_4, N_2O and CO_2, both separately and together, are now being calculated at Cambridge University.

The RAL and Cambridge University diurnal model

This one-dimensional model normally takes its starting fields of ozone temperature, etc, from the two-dimensional model, so that it reaches stability after runs of only a few model days. Alternatively the fields can be specified so that predictions can be compared with specific measurements. In this way model calculations were compared with measurements of NO and NO_2 by the Oxford radiometer.

In order to study the rapid change in NO and NO_2 at sunrise and sunset, the model's treatment of Rayleigh scattering has been improved. The model can now make accurate calculations of the error which arises in retrieving concentration profiles from solar occultation measurements if these rapid changes are ignored. In a recent paper, Roscoe and Pyle also show how this error can be calculated quite simply if these rapid changes are measured, without appeal to a model. Such measurements were made during the 1985 MAP-GLOBUS campaign, and details of the model will be further improved in collaborative comparisons with these results.

Both the two-dimensional model and the diurnal model are being used to investigate Antarctic ozone in a cooperative Cambridge University/British Antarctic Survey study.

One-dimensional models

At Harwell, predictions by a one-dimensional model were compared with measurements of the vertical profiles of 9 source gases (halocarbons, CFCs, CH_4, N_2O). Agreement was good for constituents which are removed by photolysis but not for those removed by reaction with OH.

Causes of the reduction in Antarctic ozone continue to be investigated with the BAS one-dimensional model. Many potential mechanisms can be invoked in the model but it is difficult to find a consistent explanation without more measurements.

The Meteorological Office trajectory model

Important new work at the Meteorological Office used this conceptually simple model to predict changes in constituents in parcels of air going into and out of the Arctic polar night. Results have been compared to LIMS measurements along these trajectories. Possible explanations of the reduction in Antarctic ozone are also being investigated with the model; changes in constituents after encountering polar stratospheric clouds are also being investigated.

RESEARCH RECOMMENDATIONS 13

The NASA-WMO-UNEP Assessment (NASA/WMO, 1986) contains a comprehensive list of recommendations for research in all areas of the subject. The following list has been compiled to take account not only of priority research needs, but also of existing expertise in the UK, and should not be regarded as exhaustive.

1. FIELD MEASUREMENTS

(a) Attempts should be made to measure profiles of the reservoir gases (HOCl, H_2O_2, HNO_4, N_2O_5, $ClONO_2$) in the stratosphere. In particular measurement of hydrogen peroxide should be attempted, using the sensitive fluorescence method developed at NCAR Boulder. Both aircraft and balloon platforms may be useful for these measurements.

(b) Measurements of the radicals O, OH, HO_2 and ClO, should be made more frequently and by different, independent, techniques. This argues for the maintenance of vigorous programmes of measurements from the ground, from balloons and from aircraft.

(c) Efforts should be made to improve measurement techniques to allow simultaneous measurement of several stratospheric constituents; for example O and O_3 simultaneously; NO, NO_2 and O_3 simultaneously.

(d) Instruments capable of measuring stratospheric constituents from the ground, from balloons or from aircraft, should, if at all possible, be deployed in Antarctica during the next US National Ozone Expedition.

(e) Measurements of the trends in surface concentrations of many of the source gases should continue to be made, preferably on southern hemisphere air samples collected by British Antarctic Survey.

(f) Vertical profiles of the major bromine source gases should be measured in the stratosphere.

(g) Balloons which can follow isentropic trajectories, and lightweight instruments to make measurements from them, should be developed.

2. LABORATORY MEASUREMENTS

The following kinetics studies should be undertaken:

(a) Reactions involved in the production of NO and NO_2 from N_2O.

(b) Branching ratios for HCl formation in reactions of ClO with OH and H_2O.

(c) Kinetics of reactions at very low temperatures, in particular those which are believed to proceed through long-lived intermediates. Photolysis of N_2O_5 and other temporary reservoirs at low temperatures.

(d) Reactions of BrO, the coupled BrOx - ClOx - catalysed ozone destruction cycle and reactions of higher chlorine oxides such as Cl_2O_2.

(e) Heterogenous reactions, particularly those involving temporary reservoirs such as N_2O_5 and chlorine nitrate.

There is a continuing need for increased accuracy of spectroscopic data in support of remote sensing of stratospheric constituents. In particular:

(a) More measurements are required of the line shapes of H_2O in the infra-red, particularly the far wings of lines.

(b) More measurements should be made of the line shapes of NO_2 in the infra-red, including at low temperatures.

(c) More measurements are needed of the strengths of the weaker infra-red lines of H_2O, where they interfere with measurements of other constituents.

(d) The visible spectrum of NO_2 at low temperatures should be further investigated.

3. THEORY AND MODELLING

(a) A fully interactive general circulation model, including detailed photochemistry, should be developed.

(b) Models should be developed to study the interactive effects on climate and atmospheric chemistry of the increasing concentrations of a variety of source gases (CFCs, CH_4, N_2O) which are both radiatively active and the source of stratospheric radicals.

(c) Comparative meteorologies of the northern and southern polar latitudes, and associated tracer transports, should be studied.

(d) Measurements and models should be developed to study the problems associated with those small scale features usually averaged over in current models.

(e) Trajectory models should be developed further.

ACKNOWLEDGEMENTS

This report is a review of research which has been carried out by a large number of organisations in the UK and elsewhere; whilst it is impossible to mention them all by name, the Review Group would like to acknowledge their contributions.

The Group would like to thank Dr Robert Watson, of NASA Headquarters, Washington DC, for supplying the following diagrams from the NASA-WMO-UNEP Assessment:

Figures 3.1, 3.5, 3.7, 5.5, 6.3, 6.4, 6.5, 8.1 to 8.9 and 11.6.

Thanks go to Rosemary Jenkins and Ken Smith for the design and preparation of this report.

REFERENCES

In the interests of brevity and clarity, and in consideration of the likely readership, the normal procedure of giving scientific credit by use of references has been suspended for the bulk of this report. The NASA-WMO-UNEP Assessment (NASA/WMO, 1986) can be consulted for more detail in specific areas, and this has an extensive bibliography and reference list. In general, references are given in this present report where they are not included in the Assessment.

Papers referenced in Chapter 10 as "GRLNS" appeared in the November 1986 Supplement of Geophyical Research Letters (American Geophysical Union, Washington DC, 20009, USA).

Austin J, R L Jones, T N Palmer and A F Tuck (1987) Lagrangian studies of ozone in polar latitudes. To be submitted to Nature.

Callis L C and M Natarajan (1986) Ozone and nitrogen dioxide changes in the stratosphere during 1979-84. Nature, 323, 772-777.

Chemical and Engineering News (1986) "The Changing Atmosphere: Implications for Mankind", 24 November 1986.

COMESA (1975) Report of the Committee on the Meteorological Effects of Stratospheric Aircraft. Meteorological Office, Bracknell, UK.

Crutzen P J and F Arnold (1986) Nitric acid cloud formation in the cold Antarctic stratosphere: a major cause for the springtime ozone hole. Nature, 324, 651-655.

DOE (1979) Chlorofluorocarbons and their effect on Stratospheric Ozone (Pollution Paper 15), Department of the Environment, London.

Farman J C, B G Gardiner and J D Shanklin (1985) Large losses of total ozone in Antarctica reveal seasonal ClO_x/NO_x interaction. Nature, 315, 207-210.

Farman J C (1987) Recent measurements of ozone at British Antarctic Survey stations. Phil. Trans. Roy. Soc. (in press).

Hofmann D J, J W Harder, S R Rolf and J M Rosen (1987) Balloon borne observations of the development and vertical structure of the Antarctic ozone hole in 1986. Nature, 326, 59-62.

Isaksen I (1987) Ozone perturbations studied in a two-dimensional global model with temperature feedback in the stratosphere. Unpublished Report, University of Oslo.

Jones P D, T M L Wigley and P B Wright (1986) Global temperature variations between 1861 and 1984. Nature, 322, 430-434.

Labitzke K (1986) The lower stratosphere over the polar regions in winter and spring: relation between meteorological parameters and total ozone. Submitted to Annales Geophisicae.

Lacis A, J Hansen, P Lee, T Mitchell and S Lebedeff (1981) Greenhouse effect of trace gases, 1970-1980. Geophys. Res. Letts., 8, 1035-1038.

Liljequist G H (1956) Halo phenomena and ice crystals. Norwegian-Swedish-British Antarctic Expedition, 1949-52. Scientific Results, Vol II, Part 2A, Norsk Polarinstitut, Oslo.

McElroy M B, R J Salawitch, S C Wofsy and J A Logan (1986) Reductions of Antarctic ozone due to synergistic interactions of chlorine and bromine. Nature, 321, 759-762.

McKenzie R L and P V Johnston (1984) Springtime stratospheric NO_2 in Antarctica. Geophys. Res. Lett. 11, 73-75.

Molina L T and M J Molina (1987) Production of Cl_2O_2 from the self reaction of the ClO radical. J. Phys. Chem., 91, 433-436.

NASA/WMO (1986) Atmospheric Ozone 1985. WMO Global Ozone Research and Monitoring Project, Report No 16. World Meteorological Organisation, Geneva.

New Scientist (1986) "Bird's-eye view of the holes at the poles" 6 November 1986.

NSF (1986) National Ozone Expedition statement. Press Release, 20 October 1986. National Science Foundation, Washington DC.

Noxon J F (1978) Stratospheric NO_2 in the Antarctic winter. Geophys. Res. Letts. 5, 1021-1022.

Parker D E (1986) The sensitivity of estimates of trends of global marine temperatures to limitations in geographical coverage. LRFC 14, UK Meteorological Office.

PORG (1987) Ozone in the UK. UK Photochemical Oxidants Review Group. Department of the Environment, London.

Ramanathan V, R J Cicerone, H B Singh and J T Kiehl (1985) Trace gas trends and their potential role in climate change. J. Geophys. Res., 90, 5547-5566.

SCOPE (1986) The Greenhouse Effect, Climatic Change and Ecosystems. SCOPE Report No 29 edited by: Bolin B, B Doos, J Jager and R Warrick, Wiley & Sons, Chichester, England.

Solomon S, R R Garcia, F S Rowland and D Wuebbles (1986) On the depletion of Antarctic ozone. Nature, 321, 755-758.

Stauffer B, A Neftel, H Oeschger and J Schwander (1985) CO_2 concentrations in air extracted from Greenland ice samples. In: Greenland Ice Core, Geophysical Geochemistry and the Environment. AGU Monograph No 33, American Geophysical Union, Washington DC.

Stolarski R S, A J Krueger, M R Schoeberl, R D McPeters, P A Newman and J C Alpert (1985) Nimbus 7 satellite measurements of the springtime Antarctic ozone decrease. Nature, 322, 808-811.

Stordal F and I Isaksen (1986) Ozone perturbations due to increases in N_2O, CH_4 and chlorinated source gases: 2-dimensional time dependent calculations. Submitted to Tellus.

Tung K K, M K W Ko, J M Rodriguez and N D Sze (1985) Are Antarctic variations a manifestation of dynamics or chemistry? Nature, 322, 811-813.

UNEP (1983) Co-ordinating Committee on the Ozone Layer Report of the Sixth Session. UN Environment Programme, Nairobi.

Wang W C, Y L Yung, A A Lacis, T Mo and J E Hansen (1976) Greenhouse Effects due to Man Made Perturbations of Trace Gases. Science, 194, 685-690.

ANNEX A

STRATOSPHERIC OZONE REVIEW GROUP

The Stratospheric Ozone Review Group was set up jointly by the Department of the Environment and the Meteorological Office in November 1985.

Membership

Dr A F Tuck	Meteorological Office (Chairman*)
Dr R A Cox	Harwell
Dr J Farman	British Antarctic Survey (NERC)
Dr L J Gray	Rutherford Appleton Laboratory (SERC)
Dr R L Jones	Meteorological Office
Dr A O'Neill	Meteorological Office
Dr S A Penkett	University of East Anglia
Dr J A Pyle	University of Cambridge
Dr H K Roscoe	University of Oxford

Observer

Dr J Hollies	ICI plc

DOE representative

Dr G J Jenkins	Air Quality Division

* Dr Tuck took up a post at NOAA, Boulder, in July 1986.

TERMS OF REFERENCE

1. To review our current understanding of stratospheric ozone; the threat to it from man's activities, and related topics. To recommend areas where further research, within the national effort, is needed.

2. To provide DOE with advice on scientific matters relating to policy issues concerning stratospheric ozone and related topics.

ANNEX B

ACRONYMS

Institutes, Committees

CEC Commission of the European Communities

COMESA Committee on the Meteorological Effects of Stratospheric Aviation (UK)

COSPAR Committee on Space Research

DOE Department of the Environment (UK)

JPL Jet Propulsion Laboratory (USA)

KFA Kernforschungsanlage (Julich, FRG)

MAP Middle Atmosphere Programme

MPI Max Planck Institute (Lindau, FRG)

NASA National Aeronautics and Space Administration (USA)

NOAA National Oceanic and Atmospheric Administration (USA)

PORG Photochemical Oxidants Review Group (UK)

RAL Rutherford Appleton Laboratory (UK)

STRAC Stratospheric Research Advisory Committee (UK)

UNEP United Nations Environment Programme

WMO World Meteorological Organisation

Satellite experiments

ATMOS Atmospheric Trace Molecule Spectroscopy

ERBE Earth Radiation Budget Experiment

HALOE Halogen Occultation Experiment

ISAMS Improved Stratospheric and Mesospheric Sounder

LIMS Limb Infra-red Monitor of the Stratosphere

SAGE Stratospheric Aerosol and Gas Experiment

SAMS Stratospheric and Mesospheric Sounder

SBUV Satellite Backscattered Ultra Violet

SME Solar Mesosphere Explorer

SSBUV Spacelab Satellite Backscattered Ultra Violet

SSU Stratospheric Sounder Unit

TOMS Total Ozone Measuring System

TOVS Tiros Operational Vertical Sounder

UARS Upper Atmosphere Research Satellite

Others

CFC Chlorofluorocarbon

DU Dobson Unit

GCM General Circulation Model

LIDAR Light Detection and Ranging

PAN Peroxyacetyl nitrate

SST Supersonic Transport (also, sea-surface temperature)

ANNEX C

CHEMICALS AND THEIR FORMULAE

O	atomic oxygen
O_2	molecular oxygen
O_3	ozone
N	atomic nitrogen
N_2	molecular nitrogen
H	hydrogen
Cl	chlorine
Br	bromine

SOURCE GASES

CH_4	methane
N_2O	nitrous oxide
NO	nitric oxide
NO_2	nitrogen dioxide
CO	carbon monoxide
CO_2	carbon dioxide
H_2O	water
CH_3Cl	methyl chloride
CH_3Br	methyl bromide

Gases of completely man-made origin
with their common abbreviation (not necessarily the chemical name)

$CFCl_3$	CFC 11
CF_2Cl_2	CFC 12
CHF_2Cl	CFC 22
$CF_2Cl.CFCl_2$	CFC 113
$CF_2Cl.CF_2Cl$	CFC 114
CCl_4	Carbon tetrachloride (tetrachloromethane)
CH_3CCl_3	methyl chloroform
CF_2ClBr	Halon 1211
CF_3Br	Halon 1301

TEMPORARY RESERVOIR SPECIES

(Relatively stable compounds in which amounts of ozone-depleting species can be tied up.)

$ClONO_2$ ($ClNO_3$)	chlorine nitrate
HOCl	hypochlorous acid
N_2O_5	dinitrogen pentoxide
H_2O_2	hydrogen peroxide
HO_2NO_2 (HNO_4)	peroxynitric acid

RADICALS

(Short-lived compounds which react strongly in the atmosphere.)

O	atomic oxygen
OH (HO)	hydroxyl
HO_2	hydroperoxyl
NO	nitric oxide
NO_2	nitrogen dioxide
Cl	chlorine
Br	bromine
ClO	chlorine monoxide
OClO (ClO_2)	chlorine dioxide
BrO	bromine monoxide

LONG-LIVED RESERVOIRS (SINKS)

(Compounds which act to remove ozone depleting species from the stratosphere.)

HCl	hydrochloric acid
HNO_3	nitric acid

Other chemical compounds which appear as intermediates in chemical reactions are not listed; they are almost always referred to by their formula rather than their chemical name.

Printed in the United Kingdom for Her Majesty's Stationery Office
Dd240129/8.87/C15/3936/12521